Complete Your **Pre-Summer Test**

Know your current skill level

Visit **lumoslearning.com/book**

Enter the Access Code: **PLD8**

OR

Scan the QR Code

Lumos Learning

Step Up Your Skills

Lumos Summer HeadStart, College Prep Workbook: SAT/ACT Practice, Math, Reading, Writing, Science, Online Resources & AI Tutor.

Contributing Author - John Eaton
Contributing Author - Tammie Rolf
Contributing Author - Lauren Inzelbuch
Contributing Author - Karen O Brien
Contributing Author - Donald Woods
Contributing Author - Janese Mott
Executive Producer - Mukunda Krishnaswamy
Program Director - Anirudh Agarwal
Designer and Illustrator - Sowmya R.

Updated in June 2026

ISBN 13: 979-8897131082

Printed in the United States of America

FOR SCHOOL EDITION AND PERMISSIONS, CONTACT US

LUMOS INFORMATION SERVICES, LLC

PO Box 1575, Piscataway, NJ 08855-1575
www.LumosLearning.com

Email: support@lumoslearning.com
Tel: (732) 384-0146
Fax: (866) 283-6471

Lumos Learning
Step Up Your Skills

WHY SHOULD THIS PROGRAM MATTER TO YOU?

Summer Slide is a significant challenge facing students today. It refers to the decline in academic skills and knowledge that occurs when students are not engaged in educational activities during the summer months.

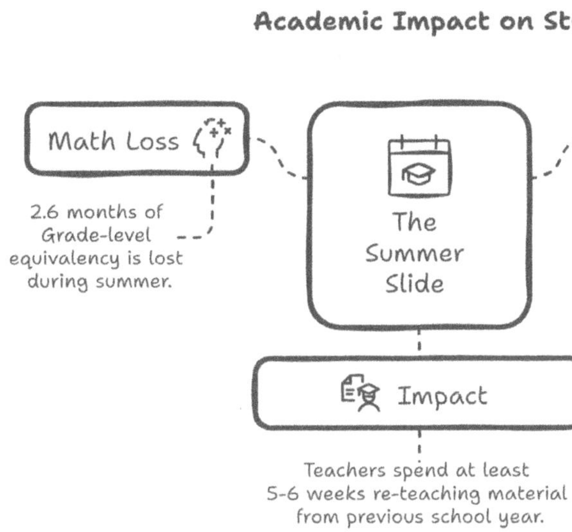

How Can You Stay Ahead?

Studies show that students who **practice over the summer retain up to 90%** of their skills, while **those who don't may lose nearly 40%** of what they learned during the school year.

The Lumos Summer Learning HeadStart Program gives you everything you need to stay ahead. It includes:

- **A Printed Workbook** with daily Math and English Practice.
- **An Online Program** with Math, English, Science, Reading, Vocabulary, and Fun Activities.

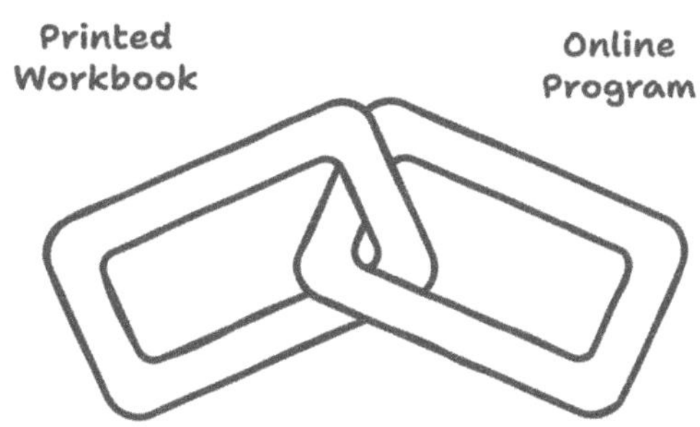

This blended learning approach ensures you stay on track, gain new skills, and start the school year with confidence. So, start your **Summer Learning HeadStart** today!

TABLE OF CONTENTS

WEEK 3 SUMMER PRACTICE

(P 47 - 69)

WEEK 4 SUMMER PRACTICE

(P 70 - 89)

WEEK 5 SUMMER PRACTICE

(P 90 - 110)

Online Activity College Readiness Mindmap | Science Activity | Weekly Essay Prompt

WEEK 6 SUMMER PRACTICE

(P 111 - 130)

Online Activity College Readiness Mindmap | Science Activity | Weekly Essay Prompt

WEEK 7 SUMMER PRACTICE

(P 131 - 149)

WEEK 8 SUMMER PRACTICE

(P 150 - 167)

Lumos Learning
Step Up Your Skills

WEEK 1
SUMMER PRACTICE

UNDERSTANDING POLYNOMIALS

DAY 1

1. Add these polynomials $(7x^3-5x^2-8x+15)+(-4x^3-2x^2+9x+6)$.

Ⓐ $3x^3-7x^2-x+21$
Ⓑ $3x^3-7x^2+x+21$
Ⓒ $11x^3-7x^2+x+21$
Ⓓ $3x^3+7x^2+x+9$

2. Subtract $(3x^3 + x^2 - 9x)$ from $(8x^3-4x^2-3)$

Ⓐ $5x^3 - 5x^2+9x+3$
Ⓑ $5x^3 - 5x^2+9x-3$
Ⓒ $5x^3 + 5x^2-9x+3$
Ⓓ $5x^3+5x^2+9x-3$

3. Multiply these polynomials $(x^2-6x+5) \times (-4x^2-x+1)$.

Ⓐ $4x^4+23x^3-13x^2-11x+5$
Ⓑ $-4x^4+23x^3-13x^2-11x+5$
Ⓒ $4x^4-23x^3-13x^2-11x+5$
Ⓓ $-4x^4+23x^3+13x^2+11x+5$

4. The expression $6a^2 - 3a + (-14a^2) - 6a + 12a^2$ is equivalent to

Ⓐ $5a^2$
Ⓑ $-5a^2$
Ⓒ $-4a^4 - 9a^2$
Ⓓ $4a^2 - 9a$

5. Which answer choice expresses the difference of $(4x^2 + 2x - 3) - (2x^2 - 5x - 1)$?

Ⓐ $2x^2 - 3x - 2$
Ⓑ $2x^2 + 7x - 2$
Ⓒ $2x^2 - 3x - 4$
Ⓓ $2x^2 + 7x - 4$

A Psalm of Life

Read the poem below and answer the questions that follow.

i. Tell me not, in mournful numbers,
Life is but an empty dream!
For the soul is dead that slumbers,
And things are not what they seem.

ii. Life is real! Life is earnest!
And the grave is not its goal;
Dust thou art, to dust returnest,
Was not spoken of the soul.

iii. Not enjoyment, and not sorrow,
Is our destined end or way;
But to act, that each to-morrow
Find us farther than to-day.

iv. Art is long, and Time is fleeting,
And our hearts, though stout and brave,
Still, like muffled drums, are beating
Funeral marches to the grave.

v. In the world's broad field of battle,
In the bivouac of Life,
Be not like dumb, driven cattle!
Be a hero in the strife!

vi. Trust no Future, howe'er pleasant!
Let the dead Past bury its dead!
Act,— act in the living Present!
Heart within, and God o'erhead!

vii. Lives of great men all remind us
We can make our lives sublime,
And, departing, leave behind us
Footprints on the sands of time;

viii. Footprints, that perhaps another,
Sailing o'er life's solemn main,
A forlorn and shipwrecked brother,
Seeing, shall take heart again.

ix. Let us, then, be up and doing,
With a heart for any fate;
Still achieving, still pursuing,
Learn to labor and to wait.

6. The tone of the poem can best be described as

Ⓐ Bored
Ⓑ Commanding
Ⓒ Tired
Ⓓ Thoughtful

7. In the first sentence, who is the narrator speaking to when he remarks "Tell me not, in mournful numbers, Life is but an empty dream!"?

Ⓐ A Psalmist
Ⓑ His friend
Ⓒ His uncle
Ⓓ No one

8. In stanza one, the line "For the soul is dead that slumbers" refers to

Ⓐ People who lie in their beds for too long.
Ⓑ People who do too much with their spare time.
Ⓒ People who do too little and are in a "sleep-like state."
Ⓓ People who live life to the fullest.

9. Whom does the narrator reference in stanza vii that should be inspirations to us?

Ⓐ great men
Ⓑ shipwrecked brother
Ⓒ the soul
Ⓓ the readers

10. True or False? In Stanza 3, the line "But to act, that each to-morrow/ Find us farther than to-day" best reflects the theme that action and growth should be defining concepts to live by

Ⓐ True
Ⓑ False

REMAINDER THEOREM

DAY 2

1. Evaluate the function $f(x) = 3x^3 - 3x^2 + 2x + 14$ at $x = 2$ using the Remainder Theorem, with synthetic division.

Ⓐ -26
Ⓑ 30
Ⓒ 26
Ⓓ -30

2. Divide $(x^3-6x^2+11x-6)$ by $(x+2)$ using synthetic division and the Remainder Theorem. What is the remainder in this division?

Ⓐ 60
Ⓑ 6
Ⓒ -60
Ⓓ 24

3. Divide (x^3+2x^2+x-7) by $(x-2)$ using synthetic division and the Remainder Theorem. What is the remainder? Choose the correct answer choice.

Ⓐ -9
Ⓑ 41
Ⓒ -19
Ⓓ 11

4. Evaluate the function $h(x) = (x^4+x^3-3x^2-4x-5)$ at $x=2$ using the Remainder Theorem, with synthetic division.

Ⓐ 1
Ⓑ 34
Ⓒ -1
Ⓓ -34

5. Find the remainder when $2x^3-5x+x-3$ is divided by $x-1$.

Ⓐ - 3
Ⓑ there is no remainder
Ⓒ - 11
Ⓓ - 5

6. Which of the following is the best example of a topic sentence?

Ⓐ There are many reasons why pollution in New York City is the worst.
Ⓑ Dogs make wonderful pets because they help you live longer.
Ⓒ Crime occurs in poverty-stricken areas.
Ⓓ Remodeling a kitchen successfully requires research and a good eye.

Read the following paragraph and answer the question.

The Dog takes care of his master's property; he follows him about, and will not let anybody hurt him. Should a stranger come, he barks, to let his master know it. He is easily taught a great many useful things. A shepherd's dog will not let the sheep go astray. It would be well if little boys and girls were always as faithful to their parents as the dog is to his master.

7. Which is the topic sentence in this paragraph?

Ⓐ The dog takes care of his master's property; he follows him about and will not let anybody hurt him.
Ⓑ A shepherd's dog will not let the sheep go astray.
Ⓒ It would be well if little boys and girls were always as faithful to their parents as the dog is to his master.
Ⓓ He is easily taught a great many useful things.

8. The syntax in a piece of writing refers to

Ⓐ Meaning of a word
Ⓑ How the words are arranged in phrases
Ⓒ How the words connect or transition between sentences
Ⓓ The implied context of the word

This type of informative writing tends to use steps. For example, if you want to bake a pie, first you need a decent pie crust. Then you fill the crust with whatever you want – cherries, blueberries, etc. Finally, you put another layer of crust on top. Then bake.

9. What is the text structure of this type of informative writing?

Ⓐ Descriptive
Ⓑ Order and sequence
Ⓒ Problem and solution
Ⓓ Cause and effect
Ⓔ None of the above

10. Which of the following transition words is most likely used to show contrast?

Ⓐ consequently
Ⓑ nonetheless
Ⓒ furthermore
Ⓓ however

ZEROS OF A POLYNOMIAL

DAY 3

1. What are the zeros of $f(x) = x^3+16x^2+64x$

Ⓐ 0, 16
Ⓑ -8, 0
Ⓒ -4, 0
Ⓓ 0, 8

2. What are the zeros of $h(x) = x^2-13x+40$

Ⓐ -5, 8
Ⓑ 5, 8
Ⓒ -8, 5
Ⓓ -8, -5

3. Find the roots of $x^2 - x - 20 = 0$.

Ⓐ {5, 4}
Ⓑ {5, -4}
Ⓒ {-5, 4}
Ⓓ {-5, -4}

4. What is the solution set of the equation $6x^2 - 18x - 18 = 6$?

Ⓐ {4, 1}
Ⓑ {-4, -1}
Ⓒ {4, -1}
Ⓓ {4, -1}

5. How many distinct roots does the equation $y = (3x - 6)^2$ have?

Ⓐ 1
Ⓑ 2
Ⓒ 3
Ⓓ 0

Twelve Years a Slave

Read the story below and answer the questions that follow.

On Christmas day, 1829, I was married to Anne Hampton, a colored girl then living in the vicinity of our residence. The ceremony was performed at Fort Edward, by Timothy Eddy, Esq., a magistrate of that town, and still a prominent citizen of the place. She had resided a long time at Sandy Hill, with Mr. Baird, proprietor of the Eagle Tavern, and also in the family of Rev. Alexander Proudfit, of Salem. This gentleman for many years had presided over the Presbyterian society at the latter place, and was widely distinguished for his learning and piety. Anne still holds in grateful remembrance the exceeding kindness and the excellent counsels of that good man. She is not able to determine the exact line of her descent, but the blood of three races mingles in her veins. It is difficult to tell whether the red, white, or black predominates. The union of them all, however, in her origin, has given her a singular but pleasing expression, such as is rarely to be seen. Though somewhat resembling, yet she cannot properly be styled a quadroon, a class to which, I have omitted to mention, my mother belonged.

6. The main purpose of this paragraph is to______.

Ⓐ Persuade
Ⓑ Describe
Ⓒ Inform
Ⓓ Defend

7. Which of the following words would best replace "inquiry?"

Ⓐ Inquisition
Ⓑ Disinterest
Ⓒ Indifference

8. The author's word choice in this piece is used to describe which type of emotion?

Ⓐ Somber
Ⓑ Reflective
Ⓒ Empathetic
Ⓓ Sentimental

Baseball

Read the story below and answer the questions that follow.

Every country has a National game. Just as cricket is the national game of England, baseball can be termed as the national game of America. Baseball became widely popular after the Civil war, and it was introduced to all parts of the country. The soldiers learned the game in their camp. Today, every village and town has its own baseball team, and the interest in the game is general. The baseball is a game for the youth, and it is not meant for men who are middle-aged, unlike golf. The chances are that if a man plays baseball in his youth, he is most likely to stay interested in the game throughout his life. Baseball requires a lot of skill. Baseball does not provide as much exercise or physical activity as tennis or football. Baseball may not always be played with a very high regard for sportsmanship, but the American public is captivated by it, and the winning of the championship series in the professional league is a celebrated national event.

Every boy knows that a baseball team consists of nine players. The various positions in the game are pitcher, catcher, first base, second base, third base, shortstop, right field, center field, and left field. The infield positions are pitcher, catcher, first base, second base, third base, and shortstop; the outfield positions are right field, center field, and left field. The position of an umpire in the baseball game is of utmost importance and his verdict on the game can result in either victory or a defeat for the playing teams. Hence, it is very important for the umpire to know the rules of the game very well and he should be impartial to both the teams. The umpire should be fair in all decisions and should stick to his decision even if the entire team opposes his decision by "kicking." Rowdyism in baseball is mainly due to this. It is important for a good player to show himself as a gentleman, whatever may be the circumstance.

Like in any other game, winning becomes the most important thing in baseball which results in a desire to score higher than the opponent becomes more important than fair play or focus on the real benefits of the game. In fact, most of the clean-cut games are played by school and college teams, and these are mostly amateurs.

9. What is the main idea of the essay?

Ⓐ How baseball compares to cricket
Ⓑ The poor behavior of baseball players
Ⓒ Why baseball is an inferior sport
Ⓓ All about baseball

10. What mistake does the writer make in paragraph 2, when she writes? Every boy knows that a baseball team consists of nine players:

Ⓐ Generalizes all boys
Ⓑ Makes baseball seem like every boy has to play it and know about it
Ⓒ All of the above
Ⓓ None of the above

CREATE EQUATIONS AND INEQUALITIES IN ONE VARIABLE

DAY 4

1. Madison is a sales associate for a transportation company. Each month she sells two cars for every 10 bicycles and four motorcycles for every car. If she makes 40 sales per month, and the variable x represents the number of cars she sells, which equation could you use to find how many cars she sells per month?

Ⓐ $x+5x+4x=40$
Ⓑ $x+5x+4x=20$
Ⓒ $2x+10x+8x=40$
Ⓓ $2x+10x+8x=20$

2. Meghan had a bag of burgers. She ate two burgers and then divided the remaining burgers amongst her friends. She has six friends and each friend received three burgers. How many burgers did Meghan originally have in her bag?

Ⓐ 14
Ⓑ 18
Ⓒ 24
Ⓓ 20

3. Pete's school supply shop has a big "Back to School" special. For $25, you get 7 folders, 5 packs of paper, 8 pens, and 15 pencils. How much does a pack of paper cost if folders cost $0.85, pens cost $1.20, and pencils cost $0.45?

Ⓐ $0.54
Ⓑ $0.45
Ⓒ $0.75
Ⓓ $0.35

4. A shopkeeper has a special where you buy four handkerchiefs, five pairs of socks, and two towels for one price. 65 specials were sold and the total sales of the specials are $2,145. What is the average price of each item in the special?

Ⓐ $11
Ⓑ $2
Ⓒ $3
Ⓓ $6

5. Which expression below is the correct representation of "the product of 4 and a number, increased by 6, is 9 less than twice the number" if n represents the variable?

Ⓐ $4n + 6 = 9 - 2n$
Ⓑ $4(n + 6) = 9 - 2n$
Ⓒ $4n + 6 = 2n - 9$
Ⓓ $4(n + 6) = 2n - 9$

6. How is a word approved for the dictionary?

Ⓐ Any word can enter the dictionary.
Ⓑ A petition needs to be started.
Ⓒ Citation-to-entry process.
Ⓓ None of the above.

7. Which sentence shows the incorrect use of the word averse?

Ⓐ My position on the topic was averse to the other candidates.
Ⓑ I was so tired that I was averse to going on a hike.
Ⓒ Mixing medications can have an averse effect on the body.
Ⓓ Why are you so averse to my involvement?

8. Which of the following factors does not influence how a word changes over time?

Ⓐ Ignorance
Ⓑ Economy
Ⓒ Cultural Environment
Ⓓ Migration

9. Which of the following is not an example of how language has changed over time?

Ⓐ Semantic
Ⓑ Lexical
Ⓒ Syntactic
Ⓓ Dictionary Use

10. Which phrase represents an adverb phrase in the following sentence: "He placed the present by the birthday cake."

Ⓐ He placed
Ⓑ the present
Ⓒ by the birthday cake
Ⓓ placed the present

1. Which equation represents the function f(x) in the table below?

x	f(x)
6	9
7	19
8	29

Ⓐ f(x)=9x-41
Ⓑ f(x)=10x-39
Ⓒ f(x)=4x+51
Ⓓ f(x)=10x-51

2. Which equation represents the function f(x) in the table below?

x	f(x)
6	19
7	15
8	11

Ⓐ f(x)=14x-43
Ⓑ f(x)=4x-33
Ⓒ f(x)=4x+53
Ⓓ f(x)=-4x+43

3. Which equation represents the function f(x) in the table below?

x	f(x)
24	11
25	16
26	21

Ⓐ f(x)=-5x+109
Ⓑ f(x)=5x-109
Ⓒ f(x)=-10x+67
Ⓓ f(x)=5x-112

4. Which equation demonstrates a direct variation of x and y when x = 4 and y = 20?

Ⓐ $y = x + 16$
Ⓑ $y = 5x$
Ⓒ $x = 5y$
Ⓓ $x = y - 16$

5. A hot air balloon leaves the ground and rises at a steady rate of 3 kilometers (k) per minute (m). Which equation below represents this relationship?

Ⓐ $k = 3 + m$
Ⓑ $k = 3m$
Ⓒ $m = 3k$
Ⓓ $m = k + 3$

Nature Study

Read the passage below and answer the questions that follow.

What is a true naturalist?—How to start a collection—Moth collecting—The Herbarium. There is no need to read big books or listen to dry lectures to study nature. In any square foot that you may pick out at random in your lawn you will find something interesting if you will look for it. Some tiny bug will be crawling around in its little world, not aimlessly but with some definite purpose in view. There is nothing in the world that will bring more pleasure into the life of a boy or girl than to cultivate a love for nature. It is one of the joys of life that is as free as the air we breathe. A nature student need never be lonely or at a loss for friends or companions. The birds and the bugs are his acquaintances. Whenever he goes afield there is something new or interesting to see and to observe. He finds— "—— tongues in trees, books in the running brooks, Sermons in stones and good in everything."

To love nature and her mysteries does not necessarily mean to be some kind of a queer creature running around with a butterfly net or an insect box. A true naturalist is simply a man or boy who keeps his eyes and ears open. He will soon find that nature is ready to tell him many secrets. After a time, the smell of the woods, the chirp of a cricket and the rustling of the wind in the pines become his pleasures.

The reason that people do not as a rule know more about nature is simply because their minds are too full of other things. They fail to cultivate the power of accurate observation, which is the most important thing of all. A practical start in nature study is to go out some dewy morning and study the first spider web you come across, noting how wonderfully this little creature makes a net to catch its food just as we make nets to catch fish, how the web is braced with tiny guy ropes to keep the wind from blowing it away in a way similar to the method an engineer would use in securing a derrick or a tall chimney. When a fly or bug happens to become entangled in its meshes, the spider will dart out quickly from its hiding place and if the fly is making a violent struggle for life will soon spin a ribbon-like web around it which will hold it secure, just as we might attempt to secure a prisoner or wild animal that was trying to make its escape, by binding it with ropes. A spider makes a very interesting pet and the surest way to overcome the fear that many people have of spiders is to know more about them.

To this insect the blades of grass are almost like mighty trees and the imprint of your heel in the ground may seem like a valley between mountains. To get an adequate idea of the myriads of insects that people the fields, we should select a summer day just as the sun is about to set. The reflection of its waning rays on their wings will show countless thousands of flying creatures in places where, if we did not take the trouble to observe, we might think there were none.

There is one very important side to nature that must not be overlooked. It consists in knowing that we shall find a thousand things that we cannot explain to one that we fully understand. Education of any kind consists more in knowing when to say "I don't know, and no one else knows either" than to attempt a foolish explanation of an unexplainable thing.

If you ask "why a cat has whiskers," or why and how they make a purring noise when they are pleased and wag their tails when they are angry, while a dog wags his tail to show pleasure, the wisest man cannot answer your question. A teacher once asked a boy about a cat's whiskers and he said they were to keep her from trying to get her body through a hole that would not admit her head without touching her whiskers. No one can explain satisfactorily why the sap runs up in a tree and by some chemical process carries from the earth the right elements to make leaves, blossoms or fruit

Nature study is not "why?" It is "how." We all learn in everyday life how a hen will take care of a brood of chicks or how a bee will go from blossom to blossom to sip honey. Would it not also be interesting to see how a little bug the size of a pinhead will burrow into the stem of an oak leaf and how the tree will grow a house around him that will be totally unlike the rest of the branches or leaves. That is an "oak gall." If you carefully cut a green one open you will find the bug in the centre or in the case of a dried one that we often find on the ground, we can see the tiny hole where he has crawled out. Did you ever know that some kinds of ants will wage war on other kinds and make slaves of the prisoners just as our ancestors did in the olden times with human beings? Did you ever see a play-ground where the ants have their recreation just as we have ball fields and dancing halls? Did you ever hear of a colony of ants keeping a cow? It is a well-known fact that they do, and they will take their cow out to pasture and bring it in and milk it and then lock it up for the night just as you might do if you were a farm boy. The "ants' cow" is a species of insect called "aphis" that secretes from its food a sweet kind of fluid called "honey dew."

The ten thousand things that we can learn in nature could no more be covered in a chapter in this book than the same space could cover a history of the world. I have two large books devoted to the discussion of a single kind of flower, the "orchid." It is estimated that there are about two hundred thousand kinds of flowers, so for this subject alone, we should need a bookshelf over a mile long. This is not stated to discourage anyone, for of course no one can learn all there is to know about any subject. Most people are content not to learn anything or even see anything that is not a part of their daily life.

6. This passage is mainly about

Ⓐ people can experience things as a source of learning.
Ⓑ how to experience nature.
Ⓒ what makes a true naturalist.
Ⓓ the growth of insects.

7. This passage could have been titled

Ⓐ "The Naturalist"
Ⓑ "Things We Cannot Explain"
Ⓒ "Learn through experience"
Ⓓ "Insect Life"

8. "Whenever he goes afield there is something new or interesting to see and to observe." In this sentence afield refers to:

Ⓐ wander
Ⓑ wrong
Ⓒ on track
Ⓓ dangerous

9. Which statement is NOT typically part of the writing revision process?

Ⓐ If writing about a quotation, break it down into smaller parts to find meaning.
Ⓑ Include specific examples with lots of details.
Ⓒ Define individual words to enhance understanding.
Ⓓ Keep sentence structure similar within a given paragraph.

10. Which phrase represents an adverb clause in the following sentence: "You can start the oven while I finish making the dough."

Ⓐ You can start
Ⓑ start the oven
Ⓒ making the dough
Ⓓ while I finish making the dough

DAY 5

CHALLENGE YOURSELF!

✓ Create Equations in Two or More Variables
✓ Reading-Writing-Language

www.lumoslearning.com/a/dc10-5

See the first page for Signup details

WEEK 1 - ESSAY PROMPT

✓ Challenging a Belief

www.lumoslearning.com/a/wc10-w1

See the first page for Signup details

7 Hacks for an Effective and Efficient Study Plan for SAT/ACT

Study Plan Tips

1. Set the duration
Is your SAT in 6 months, 3 months, or 1 month? The amount of time you have to revise will make a big impact on what your study plan looks like. Of course, the earlier you start prepping the better, but anything you study 6 months out will still need to be refreshed closer to test day (our brains have a sneaky habit of forgetting things!)

2. Take a diagnostic test
Understanding your weaknesses (and strengths!) is an important step in knowing what area to focus your study on. Taking a diagnostic test, like the one available at Lumos Learning, will give you a clear indication of your current level, and what areas need extra attention in your revision plan.

3. Seek out resources and study tools
You are not in this alone, that's for sure! There are many resources available online to take the guesswork out of SAT prep. The StepUp program from Lumos Learning breaks study into manageable chunks, with proven methods to help that info stick.

4. Take practice tests
To fully prepare for the SAT, taking full practice tests is crucial. This way, you will get used to working under pressure and a time limit, and become familiar with specific question styles. Practice tests don't have to come at the end of your study plan, either. Spread them throughout your revision and use them to constantly review your progress.

Avoid Study Time-Wasters

5. Get enough sleep
While it may feel efficient to stay up all night, those extra hours won't necessarily result in better results. You are human, after all, and your brain thrives on being looked after. Creating (and sticking to!) a sleep schedule is a proven way to help yourself out.

6. Don't just re-copy notes
When students are preparing for the SAT, it can be tempting to copy down your study material over and over in the hopes that it will stick in your memory. Unfortunately, this is not a very effective study technique and is ultimately a waste of time. Other techniques like practice questions or creating flashcards are much better at creating lasting memory pathways in your brain.

7. Avoid cramming
Cramming can be a result of either not preparing enough for your SAT, or not trusting the preparation that you have done. Either way, cramming in the week or even the night before the test is not going to effectively boost your result the way you might hope. Yes, it is important to revise your notes, but not in the desperate way that cramming implies.

COLLEGE READINESS MIND MAPS

✓ 9 Essential College Readiness Skills

www.lumoslearning.com/a/crm1

See the first page for Signup details

WEEK 2
SUMMER PRACTICE

REPRESENT CONSTRAINTS WITH EQUATIONS OR INEQUALITIES

DAY 1

1. Hillary is planning an expensive surprise party. She is choosing between two venues. The banquet hall at Hotel A will cost $431 to rent the room and $38 per guest. The banquet hall at Hotel B will cost $747 to rent the room and $16 per guest. How many people does Hillary have to invite before the cost of having the party at Hotel B is less than the cost of having a party at Hotel A?

Ⓐ 15
Ⓑ 13
Ⓒ 14
Ⓓ 16

2. The student council is holding a dance for the school. The council pays $420 to rent a local firehouse for the evening. The council must also pay $2 per guest that attends for insurance. If the council charges $8 per ticket for the dance, how many tickets must they sell in order to recover all of the costs for the facility?

Ⓐ 60
Ⓑ 70
Ⓒ 53
Ⓓ 52

3. Drake and Heidi meet in the keyboarding lab to type their term paper assignments. Drake can type at a speed of 45 words per minute and Heidi can type at a speed of 60 words per minute. What are the combined minutes of typing they will have to do if Drake's term paper is 2,160 words and Heidi's term paper is 2,640 words?

Ⓐ 44
Ⓑ 92
Ⓒ 48
Ⓓ 46

4. Zuhalie needs to rent a racing bike. She is choosing between two rental options. Option A is $50 per month with a $200 down payment. Option B is $25 per month with a down payment of $500. How many months would Zuhalie have to rent the bikes in order for the total cost of each option to be the same?

Ⓐ 24
Ⓑ 10
Ⓒ 12
Ⓓ 6

5. The ratio of staff to guests at a gala was 3 to 5. If there were a total of 576 people in the ballroom, how many guests were at the gala?

Ⓐ 216
Ⓑ 360
Ⓒ 300
Ⓓ 276

The Birthmark

Read the story below and answer the questions that follow.

A long time ago, there lived a skillful scientist who had experienced a spiritual reaction more striking than any chemical one. He had left his laboratory in the care of his assistant, washed the chemicals from his hands and asked a beautiful woman to become his wife. In those days new scientific discoveries such as electricity seemed to open paths into the area of miracles. It was not unusual for the love of science to compete with the love of a woman.

The scientist's name was Aylmer. He had so totally given himself to scientific studies that he could not be weakened by a second love. His love for his young wife could only be the stronger of the two if it could link itself with his love of science. Such a union did take place with truly remarkable results. But one day, very soon after their marriage, Aylmer looked at his wife with a troubled expression.

"Georgiana," he said, "have you ever considered that the mark upon your cheek might be removed?"

"No," she said, smiling. But seeing the seriousness of his question, she said, "The mark has so often been called a charm that I was simple enough to imagine it might be so."

"On another face it might," answered her husband, "but not on yours. No dear, Nature made you so perfectly that this small defect shocks me as being a sign of earthly imperfection."

"Shocks you!" cried Georgiana, deeply hurt. Her face reddened and she burst into tears. "Then why did you marry me? You cannot love what shocks you!"

We must explain that in the center of Georgiana's left cheek there was a mark, deep in her skin. The mark was usually a deep red color. When Georgiana blushed, the mark became less visible. But when she turned pale, there was the mark, like a red stain upon snow.

The birthmark would come and go with the emotions in her heart. The mark was shaped like a very small human hand. Georgiana's past lovers used to say that the hand of a magical fairy had touched her face when she was born. Many a gentleman would have risked his life for the honor of kissing that mysterious hand. But other people had different opinions. Some women said the red hand quite destroyed the effect of Georgiana's beauty. Male observers who did not praise the mark simply wished it away so that they did not see it. After his marriage, Aylmer discovered that this was the case with himself.

Had Georgiana been less beautiful, he might have felt his love increased by the prettiness of that little hand. But because she was otherwise so perfect, he found the mark had become unbearable.

During a period that should have been their happiest, Aylmer could only think of this disastrous subject. With the morning light, Aylmer opened his eyes upon his wife's face and recognized the sign of imperfection. When they sat together in the evening near the fire, he would look at the mark.

Georgiana soon began to fear his look. His

expression would make her face go pale. And the birthmark would stand out like a red jewel on white stone.

“Do you remember, dear Aylmer, about the dream you had last night about this hateful mark?” she asked with a weak smile.

“None! None whatever!” answered Aylmer, surprised.

The mind is in a sad state when sleep cannot control its ghosts and allows them to break free with their secrets. Aylmer now remembered his dream. He had imagined himself with his assistant Aminadab trying to remove the birthmark with an operation. But the deeper his knife went, the deeper the small hand sank until it had caught hold of Georgiana’s heart.

Aylmer felt guilty remembering the dream.

“Aylmer,” said Georgiana, “I do not know what the cost would be to both of us to remove this birthmark.
Removing it could deform my face or damage my health.”

“Dearest Georgiana, I have spent much thought on the subject,” said Aylmer. “I am sure it can be removed.”

“Then let the attempt be made at any risk,” said Georgiana. “Life is not worth living while this hateful mark makes me the object of your horror. You have deep science and have made great discoveries. Remove this little mark for the sake of your peace and my own.”

“Dearest wife,” cried Aylmer. ”Do not doubt my power. I am ready to make this cheek as perfect as its pair.”

Her husband gently kissed her right cheek, the one without the red hand.

The next day the couple went to Aylmer’s laboratory where he had made all his famous discoveries. Georgiana would live in a beautiful room he had prepared nearby, while he worked tirelessly in his lab. One by one, Aylmer tried a series of powerful experiments on his wife. But the mark remained.

Georgiana waited in her room. She read through his notebooks of scientific observations. She could not help see that many of his experiments had ended in failure. She decided to see for herself the scientist at work. The first thing that struck Georgiana when entering the laboratory was the hot furnace. From the amount of soot above it, it seemed to have been burning for ages. She saw machines, tubes, cylinders and other containers for chemical experiments. What most drew her attention was Aylmer himself. He was nervous and pale as death as he worked on preparing a liquid. Georgiana realized that her husband had been hiding his tension and fear. “Think not so little of me that you cannot be honest about the risks we are taking,” she said. “I will drink whatever you make for me, even if it is a poison.”

“My dear, nothing shall be hidden,” Aylmer said. “I have already given you chemicals powerful enough to change your entire physical system. Only one thing remains to be tried and if that fails, we are ruined!”

He led her back to her room where she waited once more, alone with her thoughts. She hoped that for just one moment she could satisfy her husband’s highest ideals. But she realized then that his mind would forever be on the march, always requiring something newer, better and more perfect.

Hours later, Aylmer returned carrying a crystal glass with a colorless liquid. “The chemical process went perfectly,” he said. “Unless all my science has tricked me, it cannot fail.”

To test the liquid, he placed a drop in the soil

of a dying flower growing in a pot in the room. In a few moments, the plant became healthy and green once more. "I do not need proof," Georgiana said quietly. "Give me the glass. I am happy to put my life in your hands." She drank the liquid and immediately fell asleep.

Aylmer sat next to his wife, observing her and taking notes. He noted everything -- her breathing, the movement of an eyelid. He stared at the birthmark. And slowly, with every breath that came and went, it lost some of its brightness.

"By Heaven! It is nearly gone," said Aylmer. "Success! Success!"

He opened the window coverings to see her face in daylight. She was so pale. Georgiana opened her eyes and looked into the mirror her husband held. She tried to smile as she saw the barely visible mark.

"My poor Aylmer," she said gently. "You have aimed so high. With so high and pure a feeling, you have rejected the best the Earth could offer. I am dying, dearest."

It was true. The hand on her face had been her link to life. As the last trace of color disappeared from her cheek, she gave her last breath.

Blinded by a meaningless imperfection and an impossible goal, Aylmer had thrown away her life and with it his chance for happiness. In trying to improve his lovely wife, he had failed to realize she had been perfect all along.

"The Birthmark" was written by Nathaniel Hawthorne. It was adapted and produced by Dana De-mange.

6. What does the birthmark on Georgiana's face symbolize?

Ⓐ The imperfection and mortality of all people
Ⓑ The beauty and grace of Georgiana
Ⓒ The presence of illness or disease
Ⓓ The necessity of flaws in humans

7. Which characteristic BEST describes Alymer's dominant trait?

Ⓐ Funny
Ⓑ Depressed
Ⓒ Lonely
Ⓓ Obsessive

8. How does Alymer change over the course of the text?

Ⓐ He is still completely obsessed with science and is unbothered by Georgiana's demise
Ⓑ He is depressed at the beginning but finds a wife and becomes happy
Ⓒ He realizes that he can never love another person as much as his work
Ⓓ He realizes that his obsession led to Georgiana's untimely death

9. The closing line of this abridged copy of the story is: "In trying to improve his lovely wife, he had failed to realize she had been perfect all along." Which story element is BEST emphasized in this line?

Ⓐ Theme
Ⓑ Conflict
Ⓒ Character development
Ⓓ All of the above

10. How does Alymer's reaction to the birthmark differ from the other men who have known Georgina?

Ⓐ The other men who knew Georgina found it 'magical'.
Ⓑ The other men are disgusted by it too.
Ⓒ Alymer finds it beautiful and mysterious, but the other men don't.
Ⓓ The other men think it is a sign that no one should marry Georgina.

1. The formula for the area of a triangle is $A=\frac{1}{2}(bh)$. Solve for b.

Ⓐ $b=\frac{A}{2h}$

Ⓑ $b=\frac{2A}{h}$

Ⓒ $b=2Ah$

Ⓓ $b=\frac{Ah}{2}$

2. The Volume of a cylinder can be expressed as $V=\pi r^2 h$. Solve for h.

Ⓐ $h=\frac{V}{\pi\sqrt{r}}$

Ⓑ $h=Vr^2$

Ⓒ $h=\frac{V}{\pi r^2}$

Ⓓ $h=V-\pi r^2$

3. Ohm's Law states $P=\frac{E^2}{R}$. Solve for E.

Ⓐ $E=PR^2$

Ⓑ $E=PR$

Ⓒ $E=P^2R$

Ⓓ $E=\sqrt{PR}$

4. The Pythagorean Theorem is $a^2+b^2=c^2$ Solve for a.

Ⓐ $a=\sqrt{c^2-b^2}$

Ⓑ $a=\sqrt{b^2-c^2}$

Ⓒ $a=c^2-b^2$

Ⓓ $a=b^2-c^2$

5. Which equation below is the formula for the circumference (C) of a circle expressed in terms of r?
$C=2\pi r$

Ⓐ $r=2C\pi$

Ⓑ $r=\frac{2C}{\pi}$

Ⓒ $r=\frac{C\pi}{2}$

Ⓓ $r=\frac{C}{2\pi}$

6. Which of these is NOT a technique used in narrative writing?

Ⓐ Dialogue
Ⓑ Pacing
Ⓒ Direct quotes
Ⓓ Multiple plot lines

7. Which type of narrator uses "I" to tell their story?

Ⓐ Second person
Ⓑ Third person limited
Ⓒ First person
Ⓓ Third person omniscient

8. In fiction, which type of character tends to grow and learn from the conflict?

Ⓐ Flat
Ⓑ Static
Ⓒ Dynamic
Ⓓ Antagonist

9. When a character speaks in a story, which type of punctuation lets the reader know that a character is speaking?

Ⓐ Periods
Ⓑ Quotation marks
Ⓒ Exclamation marks
Ⓓ Question marks

10. The correct order for a traditional story is

Ⓐ Exposition, rising action, climax, falling action, and resolution
Ⓑ Rising action, exposition, falling action, climax, and resolution
Ⓒ Exposition, rising action, resolution, climax, and falling action
Ⓓ Climax, rising action, falling action, climax, resolution

1. A report shows that a high school with 400 students, has 40% girls. The report says that 20% of the students in the school wear glasses, and that 10% of the boys in the school wear glasses. How many boys wear glasses?

Ⓐ 24
Ⓑ 240
Ⓒ 16
Ⓓ 160

2. A study found that in a small town, 40% of all households have a pet and 35% of all households have children living at home. The report also found that 60% of all households with children living at home have a pet. If a household in the town is selected at random, what is the probability that the household has no children living at home but has a pet?

Ⓐ 24%
Ⓑ 19.5%
Ⓒ 26%
Ⓓ 40%

3. Find P(W|X).

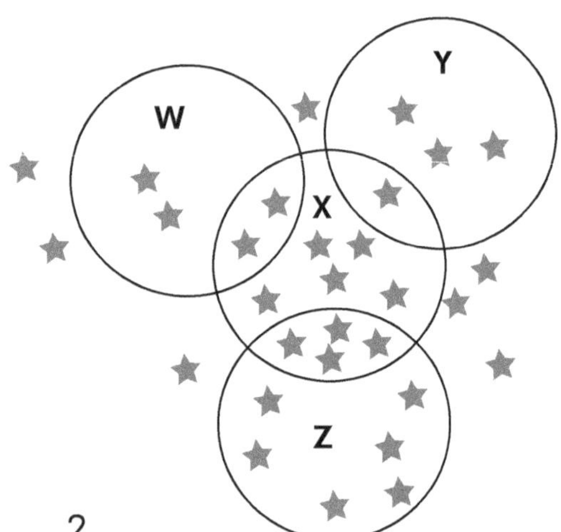

Ⓐ $\frac{2}{7}$
Ⓑ $\frac{1}{6}$
Ⓒ $\frac{1}{3}$
Ⓓ None of these

Please refer to the image below for questions 4 and 5.

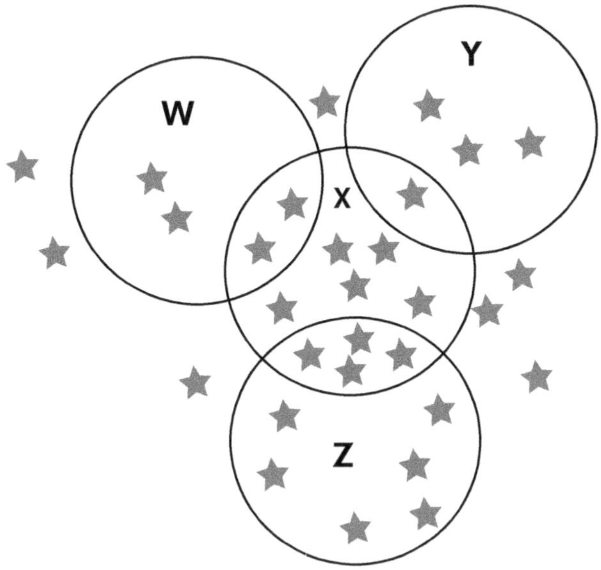

4. Find P((Y or Z)|X).

Ⓐ $\frac{1}{12}$
Ⓑ $\frac{1}{3}$
Ⓒ $\frac{5}{12}$
Ⓓ None of these

5. Find P((X or Y)|Z).

Ⓐ $\frac{1}{3}$
Ⓑ $\frac{1}{2}$
Ⓒ $\frac{1}{10}$
Ⓓ None of these

Diversity among Plants and Animals

Read the passage below and answer the questions that follow.

I know nothing about plants and animals, and my writing is based on my observations. When we compare the individuals of the same variety or sub-variety of our older cultivated plants and animals, one of the first points which strikes us is, that they generally differ more from each other than do the individuals of any one species or variety in a state of nature. And if we reflect on the vast diversity of the plants and animals which have been cultivated, and which have varied during all ages under the most different climates and treatment, we are driven to conclude that this great variability is due to our domestic productions having been raised under conditions of life not so uniform as, and somewhat different from, those to which the parent species had been exposed under nature. There is, also, some probability in the view propounded by Andrew Knight, that this variability may be partly connected with excess of food. It seems clear that organicbeings must be exposed during several generations to new conditions to cause any great amount of variation; and that, when the organization has once begun to vary, it generally continues varying for many generations. No case is on record of a variable organism ceasing to vary under cultivation. Our oldest cultivated plants, such as wheat, still yield new varieties: our oldest domesticated animals are still capable of rapid improvement or modification. As far as I am able to judge, after long attending to the subject, the conditions of life appear to act in two ways—directly on the whole organizationor on certain parts alone and indirectly by affecting the reproductive system. With respect to the direct action, we must bear in mind that in every case, as Professor Weismann has lately insisted, and as I have incidentally shown in my work on "Variation under Domestication," there are two factors: namely, the nature of the organism and the nature of the conditions. The former seems to be much the more important; for nearly similar variations sometimes arise under, as far as we can judge, dissimilar conditions; and, on the other hand, dissimilar variations arise under conditions which appear to be nearly uniform. The effects on the offspring are either definite or indefinite. They may be considered as definite when all or nearly all the offspring of individuals exposed to certain conditions during several generations are 21 modified in the same manner. It is extremely difficult to come to any conclusion in regard to the extent of the changes which have been thus definitely induced. There can, however, be little doubt about many slight changes, such as size from the amount of food, colorfrom the nature of the food, thickness of the skin and hair from climate, etc. Each of the endless variations which we see in the plumage of our fowls must have had some efficient cause; and if the same cause were to act uniformly during a long series of generations on many individuals, all probablywould be modified in the same manner. Such facts as the complex and extraordinary outgrowths which variably follow from the insertion of a minute drop of poison by a gall-producing insect, shows us what singular modifications might result in the case of plants from a chemical change in the nature of the sap.

6. What is the author's intent by stating in the opening line "I know nothing"

Ⓐ He is not a scholar of the subject matter
Ⓑ He wants the reader to draw their own opinions about what they believe
Ⓒ He is confused on the matter himself
Ⓓ He does not have evidence to support the claims

7. Why does the author assume there are endless variations between species?

Ⓐ Because of the chemical change in each species
Ⓑ All the slight changes in appearance experienced by each species
Ⓒ Species need to vary from generation to generation to survive
Ⓓ All species have unlimited points of variation

8. When reading an article, after summarizing it, the reader should always identify next:

Ⓐ Words that use a strong connotation.
Ⓑ The thesis or the main purpose of the author's work.
Ⓒ The conclusion.
Ⓓ Supporting evidence.

9. Which of the following would most be considered analysis?

Ⓐ Summarizing the text.
Ⓑ Listing the main characters.
Ⓒ Identifying the different parts of the essay – introduction, conclusion, etc.
Ⓓ Exploring how the writer supports the argument by identifying key pieces of evidence.

10. When a writer introduces a piece of evidence, he/she should always:

Ⓐ Explain how that evidence links to that point.
Ⓑ Cite where the evidence comes from.
Ⓒ Link the evidence back to the larger thesis.
Ⓓ All of the above.

WRITING FUNCTIONS THAT DESCRIBE A RELATIONSHIP BETWEEN TWO QUANTITIES

DAY 4

1. Which function correctly describes the relationship between a and b in the table below, with a as the independent variable?

a	1	2	3	4
b	4	8	12	16

Ⓐ $f(b)=4a$
Ⓑ $f(a)=\frac{a}{4}$
Ⓒ $f(a)=4a$
Ⓓ $f(b)=\frac{b}{4}$

2. Damon drove 120 miles in 2 hours and Ashley drove 135 miles in 2.3 hours. The value of the output variable of a function is always 9 more than twice the input variable. If the relationship is expressed as ordered pairs (c,d), what is the function?

Ⓐ $f(c) = 2c+9$
Ⓑ $f(d) = 2d-9$
Ⓒ $f(c) = 2c-9$
Ⓓ $f(c) = 3c+9$

3. The value of the output variable of a function is always 2 less than three times the cube of the input variable. If the relationship is expressed as ordered pairs (m,n), what is the function?

Ⓐ $f(m)=3m^3-2$
Ⓑ $f(n)=3n^3-2$
Ⓒ $f(m)=2m^3-3$
Ⓓ $f(m)=3m^3+2$

4. Which of the following functions show an exponential decay?

Ⓐ $f(x)=3(2)^x$
Ⓑ $f(x)=-5(1.4)^x$
Ⓒ $f(x)=7(0.25)^x$
Ⓓ $f(x)=0.25(1.75)^x$

5. What is the equation for your position in terms of t if you are traveling at a constant speed of v and your initial position is x_i?

Ⓐ $x(t)=x_i-vt$
Ⓑ $x(t)=x_ivt$
Ⓒ $x(t)=x_i+vt$
Ⓓ Not enough information

6. Which of the following shows the correct use of a hyphen?

Ⓐ This is my first year going into high-school.
Ⓑ I am taking a new anti-inflammatory medicine.
Ⓒ The six-year-old girl is in first grade.
Ⓓ I found a sand-dollar on the beach.

7. The verb form of "note" is

Ⓐ Notify
Ⓑ Noticeable
Ⓒ Notification
Ⓓ Notation

8. Which homophone correctly completes the following sentence:

"I wonder if ______ going to join us for dinner."

Ⓐ their
Ⓑ they're
Ⓒ there
Ⓓ thier

9. A semicolon should be placed AFTER which word in the following sentence:

"Hopefully the rain will end soon; otherwise we will have to cancel the football game."

Ⓐ Hopefully
Ⓑ soon
Ⓒ otherwise
Ⓓ cancel

10. A comma should be placed AFTER which word in the following sentence:

"Soon after moving to a new house Thomas and Denny made friends with their neighbors."

Ⓐ Soon
Ⓑ moving
Ⓒ house
Ⓓ friends

DAY 4

CHALLENGE YOURSELF!

✓ Writing Functions that Describe a Relationship Between Two Quantities
✓ Language

www.lumoslearning.com/a/dc10-9

See the first page for Signup details

CONCEPTS RELATED TO SECTORS OF A CIRCLE

DAY 5

1. Calculate the length of an arc of radius 12 ft intercepted by an angle $\theta = \frac{7\pi}{4}$

Ⓐ $\frac{21}{2}\pi$ ft
Ⓑ 21π f
Ⓒ 32π
Ⓓ 42π

2. Calculate the length of an arc of radius 16 ft intercepted by an angle $\theta = \frac{\pi}{6}$

Ⓐ $\frac{8}{3}\pi$ft
Ⓑ $\frac{16}{3}\pi$ft
Ⓒ 8πft
Ⓓ 16πft

3. Calculate the length of an arc of radius 18 ft intercepted by an angle $\theta = \frac{2\pi}{3}$

Ⓐ 3π ft
Ⓑ 6π ft
Ⓒ 12π ft
Ⓓ 24π ft

4. Calculate the area of a sector of radius 10 cm and arc $\theta = \frac{\pi}{6}$

Ⓐ $\frac{25}{3}\pi cm^2$
Ⓑ $\frac{50}{3}\pi cm^2$
Ⓒ $\frac{75}{3}\pi cm^2$
Ⓓ $\frac{100}{3}\pi cm^2$

5. Calculate the area of a sector of radius 18 cm and arc $\theta = \frac{2\pi}{3}$

Ⓐ $27\pi cm^2$
Ⓑ $54\pi cm^2$
Ⓒ $108\pi cm^2$
Ⓓ $162\pi cm^2$

Anesthesia

Read the passage below and answer the questions that follow.

The word anesthesia—means WITHOUT FEELING. This describes accurately the effect of ether in the anesthetic dosage. Although there is no pain felt during operations when anesthesia is inhaled, the nerve impulses excited by a surgical operation still reaches the brain. Not every portion of the brain is fully anesthetized, since surgical anesthesia does not kill and this allows the nerve impulses to reach the brain. This gives rise to the question: What will be the effect of trauma upon the part of the brain that remains awake? If in surgical anesthesia, the traumatic impulses cause an excitation of the wide-awake cells, how are the remainder of the cells of the brain, despite anesthesia, affected? Also, are they prevented by the anesthesia from expressing that influence of nerve stimulus in conscious perception or in muscular action? Whether the ANESTHETIZED cells are influenced or not must be determined by noting the physiologic functions of the body after anesthesia has worn off. This can be done in animals by an examination of the brain-cells. The effect of anesthesia on the vasomotor, cardiac, and respiratory centers discharging energy in response to traumatic stimuli applied to various sensitive regions of the body during surgical anesthesia has long been known. If the trauma is greater, exhaustion of the entire brain will be observed after the effect of the anesthesia has worn off. Despite the complete paralysis of voluntary motion and the loss of consciousness due to ether, the traumatic impulses that are known to reach the awake centers in the medulla also reach and influence every other part of the brain. The functional depression which is a consequence of the morphologic alterations seen in the brain-cells may be due to the low blood-pressure which follows excessive trauma as shown by the experiments. The circulation of animals was first rendered STATIC by over-transfusion, and was controlled by a continuous blood-pressure record on a drum. In each of the instances, morphologic changes in the cells of all parts of the brain were found, but it required much more trauma to produce brain-cell changesin animals whose blood-pressure was kept at the normal level than in the animals whose blood-pressure was allowed to take a downward course. In the cortex and in the cerebellum, the changes in the brain-cells were in every instance more marked than in the medulla.

6. What is the structure of this text?

- Ⓐ Compare/contrast
- Ⓑ Problem/Solution
- Ⓒ Description
- Ⓓ Cause/Effect

7. What genre does this text best fit into?

Ⓐ Narrative
Ⓑ Informative
Ⓒ Reference
Ⓓ Biography

8. Which writing structures best contribute to the overall meaning of the passage?

Ⓐ Use of varied sentence structure
Ⓑ Capitalization of keywords
Ⓒ Incorporating questions
Ⓓ None of the above

9. When you are writing to inform or explain you should:

Ⓐ List a lot of really good facts
Ⓑ Use a lot of opinions
Ⓒ Write in first person
Ⓓ None of the above

10. Which type of clause is represented in the underlined portion of the following sentence?

"Jane's English teacher, who is also in charge of the school yearbook, is retiring next year."

Ⓐ independent clause
Ⓑ adverbial clause
Ⓒ noun clause
Ⓓ relative clause

DAY 5

CHALLENGE YOURSELF!

✓ Concepts Related to Sectors of a Circle
✓ Reading-Writing-Language

www.lumoslearning.com/a/dc10-10

See the first page for Signup details

WEEK 2 - ESSAY PROMPT

✓ Accomplishment & Growth

www.lumoslearning.com/a/wc10-w2

See the first page for Signup details

How to Make Your SAT Practice a Success

Take A Diagnostic Test

The first step to success is to identify your weaknesses. Don't worry, everyone has them! The structure of the SAT consists of Math, Critical Reading, and Writing sections. Each of these will test your different skills and requires a different study and exam-taking technique. By taking an online diagnostic test, you can identify your strengths and weaknesses right from the beginning to make sure you use your study time effectively.

Utilize Online Tools

While preparing for the SAT may feel daunting, it doesn't have to be! Online prep resources, such as Lumos Learning StepUp, take the guesswork out of planning your study program. The SAT is a structured exam, meaning your study program should be equally targeted. Included proficiency markers, tech-enhanced items, and assessment practice help to create a holistic, and effective study plan.

Take Practice Tests

It seems simple, but one of the best ways you can prepare is to take practice tests. This includes examining questions from previous years and sitting fully timed exams at home to practice working under pressure. It may seem tempting to only do this once you've finished revising, but it is actually more important to incorporate practice tests into your study regimen right from the beginning. The Lumos program, for example, encourages you to take practice tests and then provides remedial exercises to address any problem areas. This means that you are effectively tackling your weaknesses right from the beginning.

Set Yourself Up For Test Day

So you've studied hard and now test day is almost here. The night before the exam is not the time to cram; rather, have a big, healthy meal and get a good night's sleep. Pack your test supplies and ID in advance, and plan to arrive way ahead of schedule. Rest assured that all your hard work will pay off, and utilize breathing practices to try and stay cool, calm, and collected.

Read Instructions Thoroughly

In the adrenaline of test day it can be tempting to fly through on auto-pilot, however, try to consciously read through each question carefully. Especially in the reading sections, sometimes the questions give more context than you may realize. In multiple-choice questions there may be two very similar answers, so make sure to properly eliminate all the other options before choosing your answer. It seems simple, but avoiding silly mistakes can make a difference to your final mark.

Price and Payment Support

It is no lie that college is expensive, but the wide range of costs can be confusing and prohibitive. Whether you have a certain budget in mind, are looking for an estimate to figure out a student loan, or are looking to research tuition support programs, it is important to factor these in right from the beginning of your search.

Keep Track Of Time

A good technique to try in your practice tests is to not do the questions in order. By doing all the questions you find easy first, you will spend less time dwelling on the harder ones and potentially running down the clock. Once you have answered all your guaranteed points, then you can go back to thoroughly think through the more difficult questions. Of course, keeping an eye on the time is crucial to make sure you finish each section and feel confident with your effort.

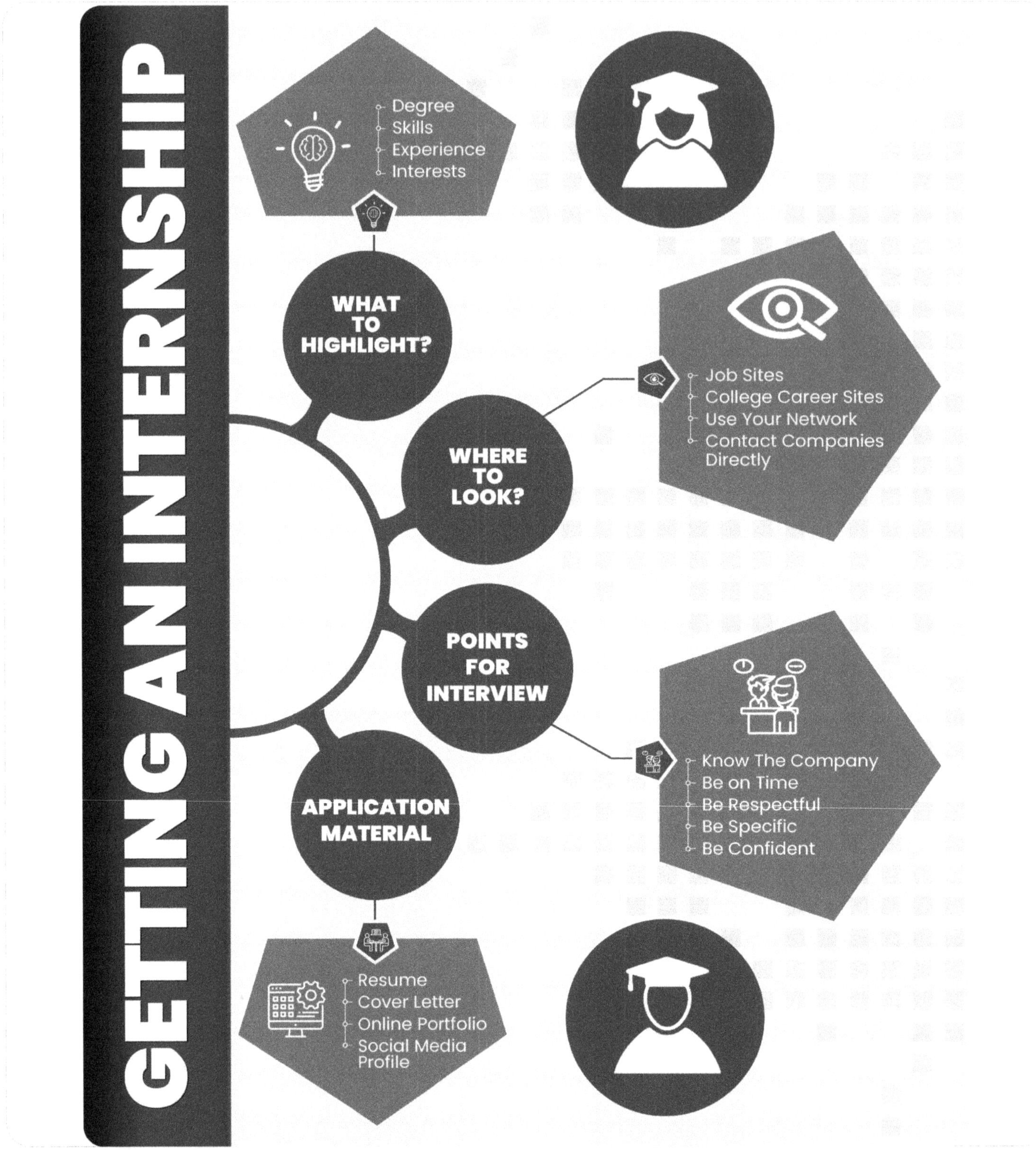

COLLEGE READINESS MIND MAPS

✓ Getting an Internship

www.lumoslearning.com/a/crm2

See the first page for Signup details

WEEK 3
SUMMER PRACTICE

SIMPLE RATIONAL AND RADICAL EQUATIONS WITH ONE VARIABLE

DAY 1

1. What is the solution to the radical equation $\sqrt{x}=0.9$?

Ⓐ 0.3
Ⓑ 0.81
Ⓒ 0.18
Ⓓ 1.8

2. What is the solution to the rational equation $\frac{2}{x+6} = \frac{-6}{x-4}$?

Ⓐ $\frac{7}{2}$
Ⓑ $\frac{-2}{7}$
Ⓒ $\frac{2}{7}$
Ⓓ $\frac{-7}{2}$

3. What is the solution to the radical equation $\sqrt{x}=0.36$?

Ⓐ 0.6
Ⓑ 0.06
Ⓒ 0.1296
Ⓓ 0.72

4. What is the solution to the rational equation $\frac{9}{(z+3)} = \frac{-3}{(z-6)}$?

Ⓐ $\frac{4}{15}$

Ⓑ $\frac{-15}{4}$

Ⓒ $\frac{-4}{15}$

Ⓓ $\frac{15}{4}$

5. Solve for a: $\sqrt{(4a)} = 12$

Ⓐ 36
Ⓑ 12
Ⓒ -12
Ⓓ -36

On Various Kinds of Thinking

by James Harvey Robinson

Read the passage below and answer the questions that follow.

The truest and most profound observations on Intelligence have in the past been made by the poet and, in recent times, by story-writers. They have been keen observers and recorders and recon freely with the emotions and sentiments. Most philosophers, on the other hand, have exhibited a grotesque ignorance of man's life and have built up systems that are elaborate and imposing, but quite unrelated to actual human affairs.

They have almost consistently neglected the actual process of thought and have set the mind off as something apart to be studied by itself. But no such mind, exempt from bodily processes, animal impulses, savage traditions, infantile impressions, conventional reactions, and traditional knowledge, ever existed, even in the case of the most abstract of metaphysicians. Kant entitled his great work A Critique of Pure Reason. But to the modern student of mind pure reason seems as mythical as the pure gold, transparent as glass, with which the celestial city is paved.

Formerly philosophers thought of mind as having to do exclusively with conscious thought. It was that within man which perceived, remembered, judged, reasoned, understood, believed, willed.

But of late it has been shown that we are unaware of a great part of what we perceive, remember, will, and infer; and that a great part of the thinking of which we are aware is determined by that of which we are not conscious. It has indeed been demonstrated that our unconscious psychic life far outruns our conscious.

This seems perfectly natural to anyone who considers the following facts:

The sharp distinction between the mind and the body is, as we shall find, a very ancient and spontaneous uncritical savage prepossession. What we think of as "mind" is so intimately associated with what we call "body" that we are coming to realize that the one cannot be understood without the other. Every thought reverberates through the body, and, on the other hand, alterations in our physical condition affect our whole attitude of mind. The insufficient elimination of the foul and decaying products of digestion may plunge us into the deep melancholy, whereas a few whiffs of nitrous oxide may exalt us to the seventh heaven of supernal knowledge and godlike complacency. And vice versa, a sudden word or thought may cause our heart to jump, check our breathing, or make our knees as water. There is a whole new literature growing up which studies the effects of our bodily secretions and our muscular tensions and their relation to our emotions and our thinking.

The term "unconscious," now so familiar to all readers of modern works on psychology, gives offense to some adherents of the past. There should, however, be no special mystery about it. It is not a new animistic abstraction, but simply a collective word to include all the physiological changes which escape our notice, all the forgotten experiences and impressions of the past which continue to influence our desires and reflections and conduct, even if we cannot remember them. What we can remember at any time is indeed an infinitesimal part of what has happened to us. We could not remember anything unless we forgot almost everything. As Bergson says, the brain is the organ of forgetfulness as well as of memory. Moreover, we tend, of course, to become oblivious to things to which we are thoroughly accustomed, for habit blinds us to their existence. So the forgotten and the habitual make up a great part of the so-called "unconscious."

If we are ever to understand man, his conduct, and reasoning, and if we aspire to learn to guide his life and his relations with his fellows more happily than heretofore, we cannot neglect the great discoveries briefly noted above. We must reconcile ourselves to the novel and revolutionary conceptions of the mind, for it is clear that the older philosophers, whose works still determine our current views, had a very superficial notion of the subject with which they dealt. But for our purposes, with due regard to what has just been said and to much that has necessarily been left unsaid (and with the indulgence of those who will at first be inclined to dissent), we shall consider mind chiefly as conscious knowledge: and intelligence, as what we know and our attitude toward it our disposition to increase our information, classify it, criticize it, and apply it.

We do not think enough about thinking, and much of our confusion is the result of current illusions in regard to it. Let us forget for the moment any impressions we may have derived from the philosophers, and see what seems to happen in ourselves. The first thing that we notice is that our thought moves with such incredible rapidity that it is almost impossible to arrest any specimen of it long enough to have a look at it. When we are offered a penny for our thoughts we always find that we have recently had so many things in mind that we can easily make a selection which will not compromise us too nakedly. On inspection, we shall find that even if we are not downright ashamed of a great part of our spontaneous thinking it is far too intimate, personal, ignoble or trivial to permit us to reveal more than a small part of it. I believe this must be true of everyone. We do not, of course, know what goes on in other people's heads. They tell us very little and we tell them very little. The spigot of speech, rarely fully opened, could never emit more than driblets of the ever-renewed hogshead of thought — noch grosser wie's Heidelberger Fass ["even larger than the Heidelberg tun"]. We find it hard to believe that other people's thoughts are as silly as our own, but they probably are.

6. Which line of the text best reflects the central idea that "People tend to accept information as true instead of trying to challenge its validity"?

Ⓐ "It is clear, in any case, that our convictions on important matters are not the result of knowledge or critical thought, nor, it may be added, are they often dictated by supposed self-interest."

Ⓑ "Most philosophers, on the other hand, have exhibited a grotesque ignorance of man's life and have built up systems that are elaborate and imposing, but quite unrelated to actual human affairs."

Ⓒ "Few of us take the pains to study the origin of our cherished convictions; indeed, we have a natural repugnance to so doing."

Ⓓ "We must reconcile ourselves to novel and revolutionary conceptions of the mind, for it is clear that the older philosophers, whose works still determine our current views, had a very superficial notion of the subject with which they dealt."

7. Which statement best reflects the central idea of the text?

Ⓐ People base their beliefs upon careful examination.
Ⓑ In the observation of human intelligence, philosophers are often at a disadvantage when compared to poets and storytellers.
Ⓒ People tend to accept information given to them as true and avoid exploring or challenging the information.
Ⓓ Most of our free time is taken up by thinking.

The Three Little Pigs

(written by Fran Havard)

Read the story below and answer the questions that follow.

It was a time of space travel, and the three little pigs were getting ready to take off. They knew they had to have the best shuttle around. There were many enemies in this distant land, and rumor has it some had very big teeth and lots of hair. The first little pig decided to build his space shuttle out of fiberglass because it was cheap and readily available. Unfortunately for him, the enemies knew exactly how to destroy that shuttle and it took one huff and puff to send that little pig crashing back to earth.

8. In the excerpt, what elements have been adapted from the classic fairy tale "The Three Little Pigs"?

Ⓐ Character
Ⓑ Theme
Ⓒ Antagonist
Ⓓ All of the Above

9. You've decided to write a short story based on the story of Hansel and Gretel but you want to change Hansel and Gretel into two orphans living in NYC in 2011. Which detail are you least likely to borrow from the original tale?

Ⓐ The breadcrumbs
Ⓑ The witch
Ⓒ The house made of candy
Ⓓ The lumberjack
Ⓔ All of the above

10. As a reader, how can you tell if something has been adapted from another text?

Ⓐ The characters' names might be similar.
Ⓑ Key quotes from the original might appear in the adaptation.
Ⓒ The theme can be the same.
Ⓓ All of the above

1. The management of a national furniture store chain wanted to know which color appealed most to potential customers. They conducted a survey of randomly selected adults in Seattle, Washington. Will this survey lead to an invalid conclusion about the most popular vegetable among the grocery chain's customers? Answers include an explanation.

Ⓐ The survey will not provide a valid conclusion because it only included adults in Seattle, Washington.
Ⓑ The survey will provide a valid conclusion because it was random.
Ⓒ The survey will not provide a valid conclusion because it was not random.
Ⓓ The survey will provide a valid conclusion because they surveyed adults.

2. A research company places signs in the open areas at a mall announcing a survey on the design of a new state flag. Below the sign is a supply of copies of the survey with a drop box for completed surveys beside it. What type of sampling is being used to conduct this survey?

Ⓐ voluntary
Ⓑ stratified
Ⓒ cluster
Ⓓ convenience

3. A group of students conduct a survey at their high school to see how many students would like lacrosse to be added as a school sport. The school has 1397 students. Of those students, 453 students play sports, and 198 are in band or chorus. The students surveyed 100 students. What is the size of the population of this survey?

Ⓐ 453
Ⓑ 1397
Ⓒ 198
Ⓓ 100

4. When should a survey be used to gather data on a population?

Ⓐ When the data cannot be gathered from a group of convenient survey participants
Ⓑ To collect a large amount of data in a relatively short amount of time
Ⓒ When every member of the population is willing to participate
Ⓓ When retired people want to make some additional income surveying people

5. The management of a national grocery store chain wanted to know which type of vegetable appealed to the most potential customers. They conducted a survey of randomly selected adults in Chicago, Illinois. Will this survey provide an invalid conclusion about the most popular vegetable of the grocery chain's customers?

Ⓐ The survey will not provide a valid conclusion because it only included adults in Chicago, Illinois.
Ⓑ The survey will provide a valid conclusion because it was random.
Ⓒ The survey will not provide a valid conclusion because it was not random.
Ⓓ The survey will provide a valid conclusion because they surveyed adults.

6. Which strategy is most effective in revising a paper?

Ⓐ Maintaining a constant work cycle with few breaks from the material.
Ⓑ Reading the paper out loud.
Ⓒ Outlining the paper after you have written chunks.
Ⓓ Skim for errors.

7. Which type of writing uses sensory details and metaphors?

Ⓐ Narrative
Ⓑ Exposition
Ⓒ Persuasive
Ⓓ Informative

8. Which of the following structures might be used in informational texts?

Ⓐ Sequence
Ⓑ Problem and Solution
Ⓒ Cause and Effect
Ⓓ All of the above

9. What is the topic of this piece of writing?

When we think about the minimum age for voting, the first question that comes to our mind is: How young is too young? The minimum ages vary in different contexts, such as for driving, marriage, joining the army, and so on. While there is an argument that 16 and 17-year-olds are too immature to vote, there is also a danger that they might not vote at all.

Ⓐ Young people
Ⓑ Voting importance for young people
Ⓒ Immaturity of 16-year-olds
Ⓓ Adults not trusting teenagers

10. What is the purpose of this piece of writing?

When we think about the minimum age for voting, the first question that comes to our mind is: How young is too young? The minimum ages vary in different contexts, such as for driving, marriage, joining the army, and so on. While there is an argument that 16 and 17-year-olds are too immature to vote, there is also a danger that they might not vote at all.

Ⓐ The purpose is to inform the reader of voting history in America.
Ⓑ The purpose is to persuade the reader to review the importance of voting with teenagers.
Ⓒ The purpose is to describe voting habits of American teens.
Ⓓ The purpose is to persuade the reader about why teenagers should have more responsibility.

1. Given the function $g(x)=x^2-2x+5$, what is the range of g(x) if the domain is {1,2,3,4}?

Ⓐ {1,2,3,4}
Ⓑ {1,2,5,10}
Ⓒ {4,5,8,13}
Ⓓ {8,13,20,29}

2. Given the function $f(x)=\frac{(x^2-x-6)}{(x-3)}$, what is the range of f(x) if the domain is {10,11,12,13,14}?

Ⓐ {12,13,14,15,16}
Ⓑ {10,11,12,13,14}
Ⓒ {16,17,18,19,20}
Ⓓ {22,33,44,55,66}

3. Given the function $g(x)=3x-23$, what is the range of g(x) if the domain is {15,18,25,27}?

Ⓐ {27,38,60,66}
Ⓑ {22,31,52,58}
Ⓒ {15,18,25,27}
Ⓓ {15,28,35,37}

4. Given the function $h(x)=4x^2-23x+8$, what is the range of h(x) if the domain is {1,3,5,7,9}?

Ⓐ {-11,-25,-7,143,225}
Ⓑ {-11,-27,-5,44,123}
Ⓒ {-11,-25,-7,43,125}
Ⓓ {7,11,25,43,125}

5. Given the function $f(x)=\frac{2x^2-13x-7}{2x+1}$, what is the range of f(x) if the domain is {2,6,10,14} ?

Ⓐ {-5, -1, 3,7}
Ⓑ {1, 3, 5, 7}
Ⓒ {-6, -2, 2, 6}
Ⓓ {-3, 1, 5, 9}

Patrick Henry Speech (March 23, 1775)

Read the story below and answer the questions that follow.

They tell us, sir, that we are weak; unable to cope with so formidable an adversary. But when shall we be stronger? Will it be the next week or the next year? Will it be when we are totally disarmed, and when a British guard shall be stationed in every house? Shall we gather strength by irresolution and inaction? Shall we acquire the means of effectual resistance by lying supinely on our backs and hugging the delusive phantom of hope, until our enemies shall have bound us hand and foot? Sir, we are not weak if we make proper use of those means which the God of nature hath placed in our power. The millions of people, armed in the holy cause of liberty, and in such a country as that which we possess, are invincible by any force which our enemy can send against us. Besides, sir, we shall not fight our battles alone. There is a just God who presides over the destinies of nations, and who will raise up friends to fight our battles for us. The battle, sir, is not to the strong alone; it is to the vigilant, the active, the brave. Besides, sir, we have no election. If we were base enough to desire it, it is now too late to retire from the contest. There is no retreat but in submission and slavery! Our chains are forged! Their clanking may be heard on the plains of Boston! The war is inevitable—and let it come! I repeat it, sir, let it come.

It is in vain, sir, to extenuate the matter. Gentlemen may cry, Peace, Peace—but there is no peace. The war is actually begun! The next gale that sweeps from the north will bring to our ears the clash of resounding arms! Our brethren are already in the field! Why stand we here idle? What is it that gentlemen wish? What would they have? Is life so dear, or peace so sweet, as to be purchased at the price of chains and slavery? Forbid it, Almighty God! I know not what course others may take; but as for me, give me liberty or give me death!

6. In the opening line of Paragraph 1, Henry uses the word "formidable" to describe his enemy, Great Britain. Based on the context of the word, what can you assume that "formidable" means?

Ⓐ Impressively large and powerful
Ⓑ Mean
Ⓒ Aggressive
Ⓓ Not worthy of one's time

7. Henry uses ____________ in the lines "Shall we acquire the means of effectual resistance by lying supinely on our backs and hugging the delusive phantom of hope" because it's clear he doesn't mean they should lie down on their backs and hang onto hope.

Ⓐ sarcasm
Ⓑ direct speech
Ⓒ a simile
Ⓓ a declarative statement

8. The series of questions that he asks the audience most ardently reflect his desire for the colonial people to be:

Ⓐ Weak and docile
Ⓑ Angry and revolutionary
Ⓒ Strong but passive
Ⓓ Brave and stoic

9. The quote, "Is life so dear, or peace so sweet, as to be purchased at the price of chains and slavery?" uses figurative language to compare freedom to remaining under British rule. Which type of figurative language does Henry use?

Ⓐ Simile
Ⓑ Metaphor
Ⓒ Onomatopoeia
Ⓓ Assonance

10. What rhetorical device does Henry use in this quote: "There is no retreat but in submission and slavery! Our chains are forged! Their clanking may be heard on the plains of Boston!"

Ⓐ Metaphor and hyperbole
Ⓑ Hyperbole and personification
Ⓒ Hyperbole only
Ⓓ A and B

1. Select the best word to fill in the blank in this sentence: In geometry, a line is defined as a set of points on a plane that extend in two directions without ______.

Ⓐ curving
Ⓑ ending
Ⓒ bending
Ⓓ moving

2. Select the best word to fill in the blank in this sentence: In geometry, a _______ is a set of points on a flat surface or a 2-dimensional object, stretching to infinity in all directions.

Ⓐ line
Ⓑ plane
Ⓒ angle
Ⓓ circle

3. Select the best word to fill in the blank in this sentence: In geometry, a circle is a set of points in a plane that are at an equal _______ from a given point called the center.

Ⓐ distance
Ⓑ angle
Ⓒ direction
Ⓓ point

4. Select the best word to fill in the blank in this sentence: In geometry, two lines that are perpendicular to each other intersect at a(n) _____ angle.

Ⓐ obtuse
Ⓑ straight
Ⓒ acute
Ⓓ right

5. Select the best word to fill in the blank in this sentence: In geometry, a solid object is a set of points that forms a figure that has _______ dimension(s).

Ⓐ one
Ⓑ two
Ⓒ three
Ⓓ four

6. Which sentence contains a misspelled word?

Ⓐ I definetly want to see the new movie that came out.
Ⓑ Apparently, I need to do some weight training for football.
Ⓒ That fall I just had was very embarrassing.
Ⓓ There is not a new car in the foreseeable future.

7. Which of the following is the adjective form of compete?

Ⓐ competing
Ⓑ competition
Ⓒ competitive
Ⓓ competed

8. Which sentence uses a colon in the correct place?

Ⓐ Please bring these items to class: pencil, textbook, and a calculator.
Ⓑ Please bring these items: to class pencil, textbook, and a calculator.
Ⓒ Please: bring these items to class pencil, textbook, and a calculator.
Ⓓ Please bring these items to class pencil textbook: and a calculator.

9. Which homophone correctly completes the following sentence: "Sometimes I feel ____ tired to get up in the morning."

Ⓐ to
Ⓑ two
Ⓒ toe
Ⓓ too

10. Which is NOT a correct use of a colon?

Ⓐ To combine two dependent clauses.
Ⓑ To introduce a list.
Ⓒ To start a formal business letter.
Ⓓ To introduce a quote longer than three lines.

1. Which line plot correctly displays the same information as the number set below?

{3,6,7,3,5,8,3,5,4,8,6,7,5,5,4,8,6}

Ⓐ 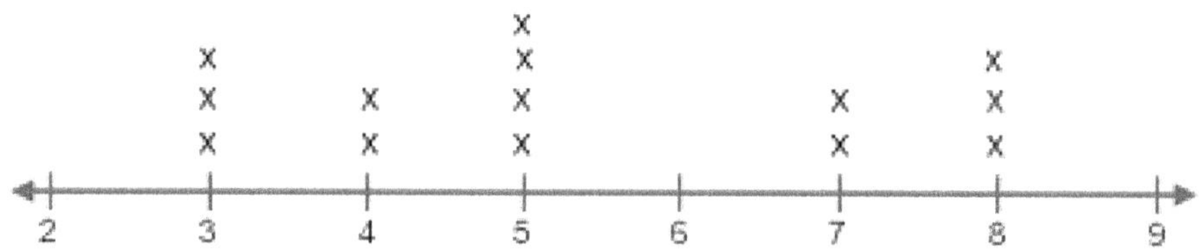

Ⓑ 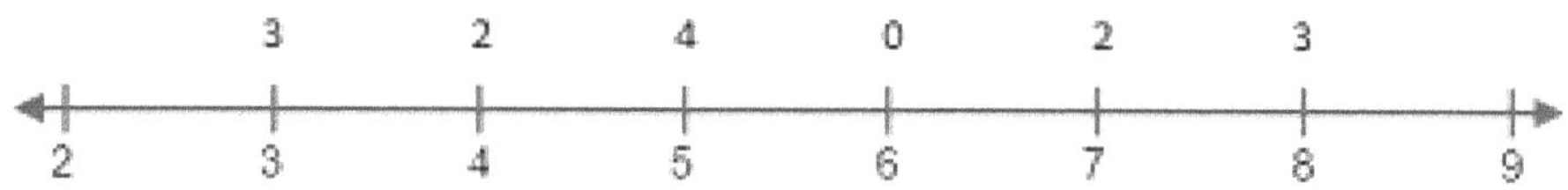

Ⓒ 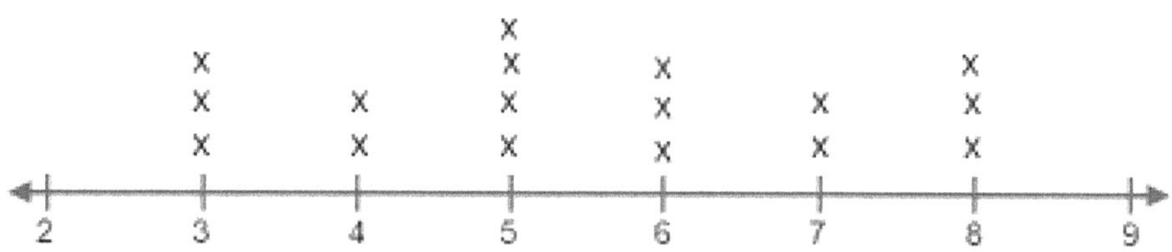

Ⓓ 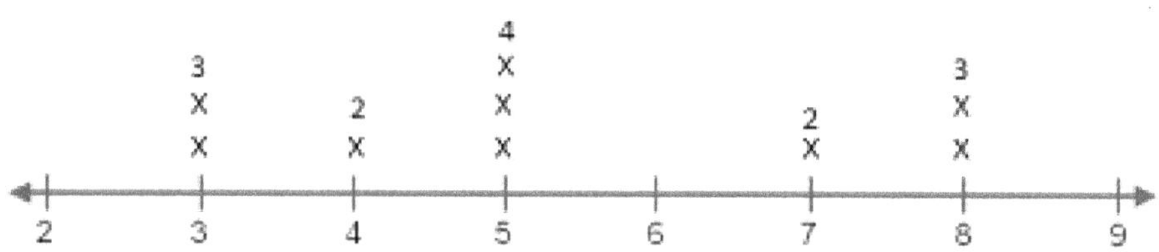

2. What is the median of the data set represented by the box and the whiskers plot below?

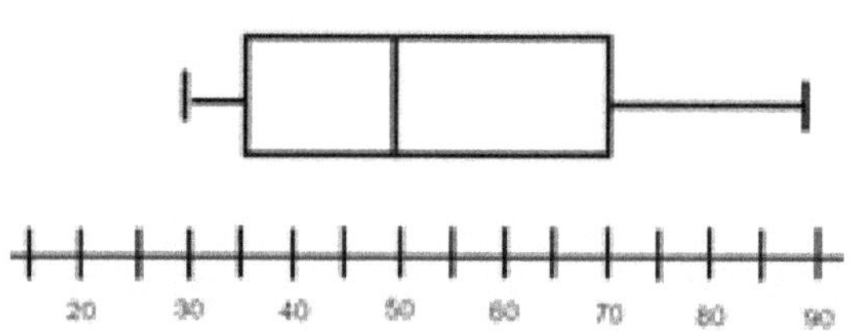

Ⓐ 30
Ⓑ 40
Ⓒ 50
Ⓓ 70

3. Which data set is displayed in the histogram shown below?

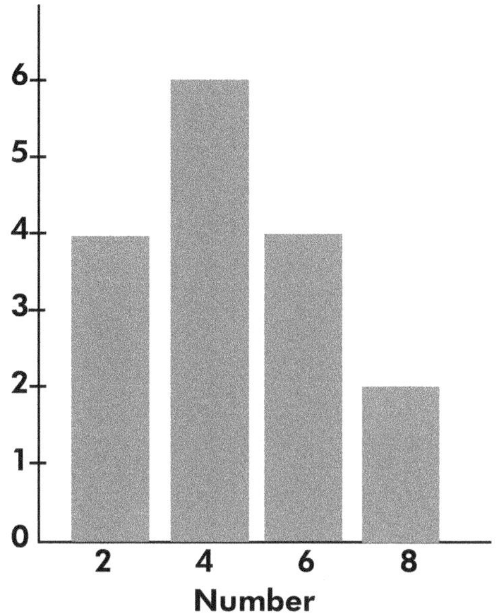

Ⓐ {2,2,4,4,2,2,6,6,6,8,2,4,6,8,6,4,2}
Ⓑ {2,4,4,2,4,6,6,8,2,4,6,8,6,4}
Ⓒ {2,4,4,4,2,2,4,6,6,8,2,4,6,8,6,4}
Ⓓ {2,4,4,4,2,2,4,6,8,2,6,8,6,4}

4. What is the first quartile of the data set represented by the box and the whiskers plot below?

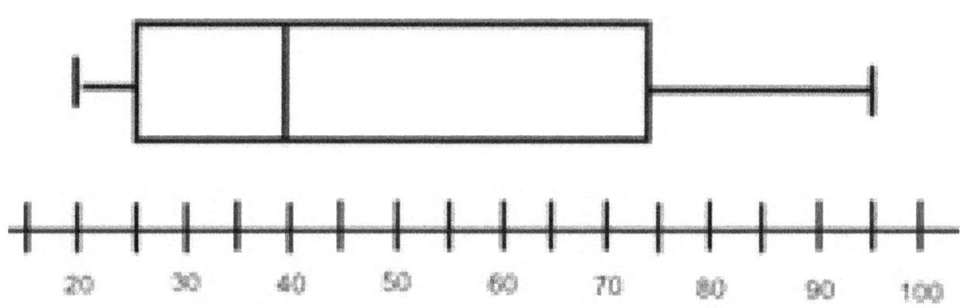

Ⓐ 25
Ⓑ 40
Ⓒ 80
Ⓓ 95

5. Below is a dot plot showing the scores students received on a quiz. Which of the following statements is true based on the dot plot?

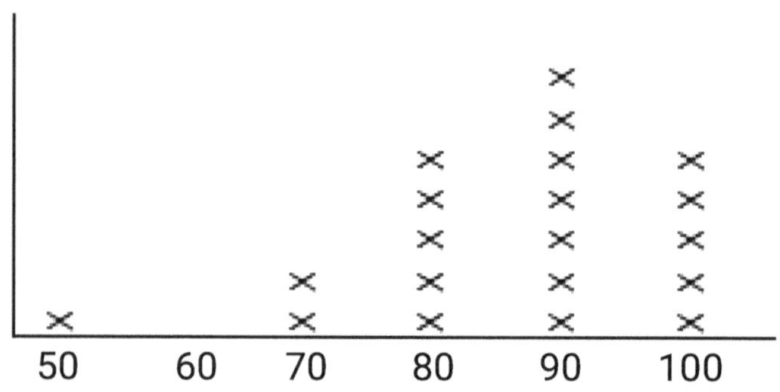

Test Grades

Ⓐ One student received a score of 60.
Ⓑ More students scored a 90 than any other score.
Ⓒ 70 was the score received by the fewest students.
Ⓓ More students scored a 70 or 80 than scored a 90.

The Exotic Plants of New Mexico

by ROSS CALVIN

Read the passage below and answer the questions that follow.

The problem of discussing plant life becomes complicated, for some of those arriving in New Mexico may be compound microscope botanists, some collectors, others pathologists, still others geneticists. Some will be chemists or mathematicians, some others untrained in the ways of growing things and mainly interested in seeing while they tour. But one thing is fairly certain- most visitors will come from a distance, so it may be useful to invite them to observe what they can readily see on the way hither, which will be a relatively painless method of amassing some information.(1) Yet the method of arrival itself suggests choices, options, and exercises in probability. Do visitors come in covered wagons, or in jet planes; by bus, car, train, or some other way? The most convenient way, doubtless, is by plane traveling six hundred miles an hour at a height of some thirty thousand feet; but the most rewarding way is by saddle or on foot as the early collectors came.(2) Since one cannot know his own country well if he knows no other, a visitor from the east arriving at an altitude of five miles will probably be more conscious than others of drastic changes in the landscape when he first looks down on New Mexico. Instead of the universal green of the east, he will note the earth as an unaccustomed tawny, reddish brown expanse, and this will be its common color through most months of the year. He will note the absence of rivers, but the presence of mountains, which generally ring the horizon. He will wonder, after traversing the border counties at the western edge of the Great Plains, at the infrequency of plowlands, and will promptly conclude that New Mexico is one vast, bare desert- an impression that will be corrected rapidly when he visits the dark National Forests where the slim, crowded spruces and firs tower skyward.

(3) The visitor from El Paso, as the Chihuahua desert (the name is fairly deserved) slides backward underneath him, will be amazed at the serpentine stripe of vivid verdure below, all completely leveled, and all marked off by geometrical lines and tightly bordered by pale hills where the vegetation is reduced to tiny dots not tall enough to cast a shadow. Still proceeding north, he will presently be above a ninety-mile stretch where the vegetation is so skimpy and the river so closely boxed in by desolate hills that Spanish explorers bestowed on it the sinister name Jornada del Muerto" Dead Man's Journey, which is self-explanatory. (4) But on the other hand, a visitor descending into New Mexico from the northland in wintertime above this same Rio Grande will traverse a featureless jumble of peaks on which the black coniferous forest no longer shows as black at all but, snow- encased, gleams as white as the naked snowfields themselves. And he will find incredible the immense range of climate spanned in a flight only an hour long. (5) The jet traveler approaching from the west beholds the greatest, most spectacular panorama of all, the true wasteland of the desert, with vast blocks uplifted from out of earth's crust, plateaus dissected by giant empty canyons whose walls are seared to utter nudity by ages of the most intense sunshine on our continent. Here the architecture of nature is all designed on a gigantic scale, and the climate exhibits influences of a mighty ocean, of

towering mountain chains, and the range of wandering planetary winds which created the desert. From the altitude of a cruising jet, the distances, the speed, the color, and the light are indescribably exhilarating.(6) Since the traveler arriving from any direction can see for identification only a few species, it will suffice to point out only two of the harbingers of eastern New Mexico, both of which are dominant on that area of the Great Plains: the blue grama grass, which in fall and winter bears the graceful sickle-shaped head, and the small Yucca glauca which cannot be mistaken for anything else.(7) Along the northern line, the visitor should look for the somber, almost black, forest of spruce and fir with its white patches of aspen on the 10,000-foot slopes. Descending farther southward, even a casual observer will note that these species give way first to the yellow pines, then to the junipers ("cedars") and dwarf pines or piñons. The visitor arriving from the south will find most conspicuous of all the olive-green creosote bush in its summer-winter foliage, next, the omnipresent mesquite (Prosopis), then the various Opuntia cacti (chollas and prickly pears), and the yuccas.(8) As he comes in from Arizona, the splendid saguaros will beckon him toward the gateway of New Mexico, but they will nowhere enter it themselves. The vase- shaped body of the ocotillo consisting of unbranching, ten-foot wands should be recognized at a glance, as likewise should the various yuccas and agaves. And in early springtime, a low tree with a rounded mass of golden bloom will joyously proclaim its identity as a paloverde.

(9) The desert plants thus enumerated can be recognized along various stretches of U.S. 66 as one holds his speed at a conservative ninety or, even better, at a conservative sixty. The amateur scientist will need to do some homework before he learns much systematic botany or plant physiology. It takes considerable study to comprehend the slow, age-long process which has resulted in the evolution of all this grotesque vegetation: the interaction of extreme aridity, high temperature, low humidity, high evaporation, infrequent precipitation, the hostility of numerous chemical solutions, and the natural rapacity of the animal population. Each plant must have an adaptation of its own, from the tiny winter annual, which must race through its course from a tender rosette to ripened seed before the searing sunshine arrives, on up through the series to the stoic saguaro, which seems able to scorn nature's enmity. (10) A final warning to the air traveler. He should be sure before deplaning that he has observed a bajada. It is a must. Bajada is one name, outwash slope is another, alluvial fan is still another. The stewardess should inquire of each student, and if his answer is "no," she should reply, "Then you've just flunked the course." A bajada is the most characteristic single feature of all in the sloping, stony topography of the desert"and remember that the surface of the desert is regularly sloping and stony. The Spanish word itself means only a down slope, but it has come to have a semi technical sense"a fan-shaped apron of rocky debris at the foot of a mountain where it has settled out of the summer torrents and there keeps accumulating in successive floods. As the mass builds up, the bigger fragments drop out closest to the parent crag, then the small ones, finally the coarse gravel, and lowest of all on the slope the fine sand. Thus the fragments are graduated in size, in distance traveled, and in the nature of soil resulting. As the fan increases in size and laps up higher and higher on the mountain's flank, the geologists often describe the result by saying that the mountain "is burying itself under its own debris." From the air, the lines of drainage will be seen to have assumed the form of a tree with its trunk downward, its branches high in air. But the most interesting aspect of a bajada is the nice proportion which nature works out between the plant species and their location on the curve. (11)

6. With which of the following statements would the author most likely agree?

Ⓐ The direction you travel from enhances your appreciation of the topography.
Ⓑ Natives of the state of New Mexico know the most about their plant life.
Ⓒ Not many travelers come to New Mexico to see their plant life.
Ⓓ Only a small number of plants are truly worth seeing.

7. How does the main idea differ from the thesis statement of a passage?

Ⓐ The thesis statement is just one piece of evidence in a passage.
Ⓑ The main idea is expressed in the thesis statement.
Ⓒ The thesis statement introduces your main idea.
Ⓓ The main idea and thesis statement are the same.

8. The event in the plot that kicks off the rising action is called:

Ⓐ Trigger incident
Ⓑ Inciting incident
Ⓒ Climax
Ⓓ None of the above

9. Which of the following is the MOST important part of informing or explaining?

Ⓐ Opinions
Ⓑ Facts
Ⓒ Hyperbole
Ⓓ None of the above

10. What punctuation should separate the dependent clause from the independent clause in the following sentence? "Starting with Rosa Parks's refusal to sit in the back of the bus the Montgomery bus boycott brought national attention to the issue of segregation."

Ⓐ semicolon (;)
Ⓑ comma (,)
Ⓒ colon (:)
Ⓓ period (.)

DAY 5

CHALLENGE YOURSELF!

✔ Representing Data in Graphs
✔ Reading-Writing-Language

www.lumoslearning.com/a/dc10-15

See the first page for Signup details

WEEK 3 - ESSAY PROMPT

✔ Share an Essay

www.lumoslearning.com/a/wc10-w3

See the first page for Signup details

Ace the SAT by Turning Disappointments Into Success

Reflect and review

When you first receive your results, you will probably feel emotional. That is perfectly okay, and you should let yourself have a big cry. Once you have released some of the initial shock and disappointment, you will be in a better place to rationally assess where you went wrong.

Take some time to go through each section of your results carefully. By understanding your weaknesses, you can work on making them your strengths. It is also worth reflecting on the actual test day experience: did your nerves get in the way? Did you run out of time? Did you have trouble sleeping the night before? Did any of the questions knock your confidence?

Mindfulness techniques can be a very important part of studying, so incorporating some into your routine will help you feel more confident next time.

Revise and reach out

If you're going to ace the SAT, it's time to create a new study plan. Now you've identified your weaknesses, you can gather resources and design a schedule that targets these areas. Now would also be a great time to reach out for help – study resources such as Lumos Learning's StepUp program make it easy to structure a simple and effective revision plan.

You've probably studied a lot of the content, so practice exams are going to be a priority this time around. With StepUp, you can revise, take a practice test, review your results and hone in on areas you're still struggling with. Don't worry, you're not in this alone – there is help out there!

Stay positive and re-prioritize

If you want to succeed, you need to believe in yourself first and foremost. Remember that you are capable and worthy, and there is so much more to you than test results. Now is a time to take a clear look at your priorities and plans for college, and assess if your goals have perhaps shifted. While you may have a dream college in mind, you may also take this time to diversify your options to relieve some of the pressure you've put on yourself.

You should be proud of yourself for making it this far, and it is that positive attitude that will take you over the finish line! Remember: there is so much more to you than an SAT score, and the irony is that once you realize that you will surely have the confidence to ace it.

SELECTION OF TOPICS

Common App Prompts
Brainstrom Ideas
Express Yourself

PRE-WRITING / OUTLINING

Free Writing
Order the Ideas
Remove Irrelevant Details

BODY OF ESSAY

Introduction
Billboard / Thesis
Body Paragraph 1
Body Paragraph 2
Conclusion

METHOD OF WRITING

Word Limit:650
5 Paragraphs
Write Several Drafts
Revise and Edit

WEEK 4
SUMMER PRACTICE

SOLVE LINEAR EQUATIONS AND INEQUALITIES WITH ONE VARIABLE

DAY 1

1. What is the solution to the equation $\frac{x}{5} - 11 = 2$?

Ⓐ $\frac{13}{5}$
Ⓑ -65
Ⓒ 56
Ⓓ 65

2. What is the solution to the inequality $5 > \frac{x+16}{7}$?

Ⓐ $x>19$
Ⓑ $x<19$
Ⓒ $x \geq 19$
Ⓓ $x \leq 19$

3. What is the solution to the equation $10x-5x-4=9$?

Ⓐ 8
Ⓑ $\frac{5}{13}$
Ⓒ $\frac{13}{5}$
Ⓓ 1

4. What is the solution to the inequality $6 < \frac{y-12}{6}$?

Ⓐ $y<48$
Ⓑ $y>48$
Ⓒ $y \leq 48$
Ⓓ $y \geq 48$

5. What is the solution to the inequality $3 \geq \frac{x+7}{8}$?

Ⓐ $x>17$
Ⓑ $x<17$
Ⓒ $x \geq 17$
Ⓓ $x \leq 17$

Freedom

Read the poem below and answer the questions that follow.

Freedom from fear is the freedom
I claim for you my motherland!
Freedom from the burden of the ages, bending your head,
breaking your back, blinding your eyes to the beckoning
call of the future;
Freedom from the shackles of slumber wherewith
you fasten yourself in night's stillness,
mistrusting the star that speaks of truth's adventurous paths;
freedom from the anarchy of destiny
whole sails are weakly yielded to the blind uncertain winds,
and the helm to a hand ever rigid and cold as death.
Freedom from the insult of dwelling in a puppet's world,
where movements are started through brainless wires,
repeated through mindless habits,
where figures wait with patience and obedience for the
master of show,
to be stirred into a mimicry of life.

6. This poem can best be described as:

Ⓐ A call to wake up and demand freedom.
Ⓑ A call to revolt against the government.
Ⓒ A plea to his people to remain stoic in the face of tyranny.
Ⓓ A comment on how lonely the people of his country are.

7. The narrator of the poem demands freedom from the following:

Ⓐ from fear
Ⓑ from shackles of slumber
Ⓒ from the anarchy of destiny
Ⓓ all of the above

8. Select whether the following statement is true or false.

The speaker of the poem is disappointed by his country.

Ⓐ True
Ⓑ False

9. Who do you think the "master of the show" is?

Ⓐ The ruling government
Ⓑ The people
Ⓒ The animals
Ⓓ The Americans

10. The point of view of the poem "Freedom" can best be described as:

Ⓐ first-person
Ⓑ second-person
Ⓒ third-person objective
Ⓓ third-person limited
Ⓔ third-person omniscient

1. What is the difference between the means of these two sets of data?

Set A: {2,4,6,8,10,12}
Set B: {3,5,7,9,11,13}

Ⓐ 7
Ⓑ 1
Ⓒ 8
Ⓓ 15

2. What is the difference in the medians of Set A and Set B, displayed in the line plot below?

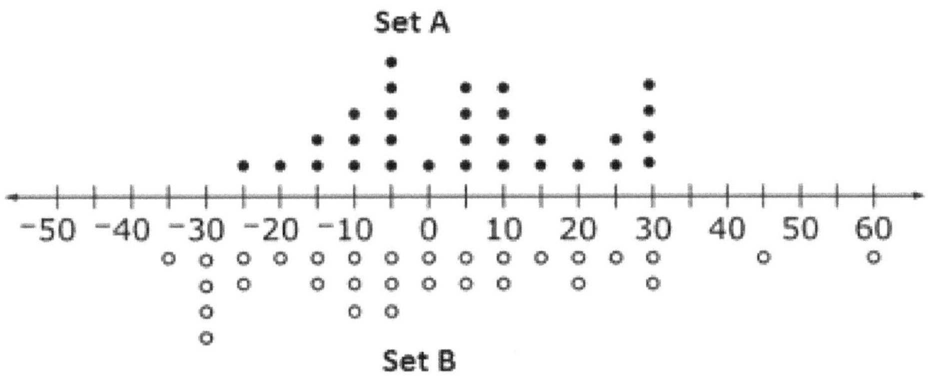

Ⓐ 5
Ⓑ 10
Ⓒ -5
Ⓓ 0

3. What is the difference between the medians of these two sets of numbers?

Set A: {6,8,10,12,14,16,18,20,22}
Set B: {3,5,7,9,11,13,15,17,19}

Ⓐ 14
Ⓑ 11
Ⓒ 3
Ⓓ 25

4. What is the difference between the medians of Set A and Set B, displayed in the box plots below?

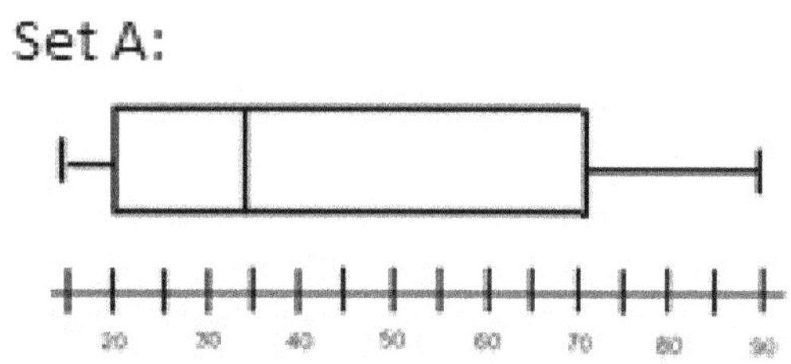

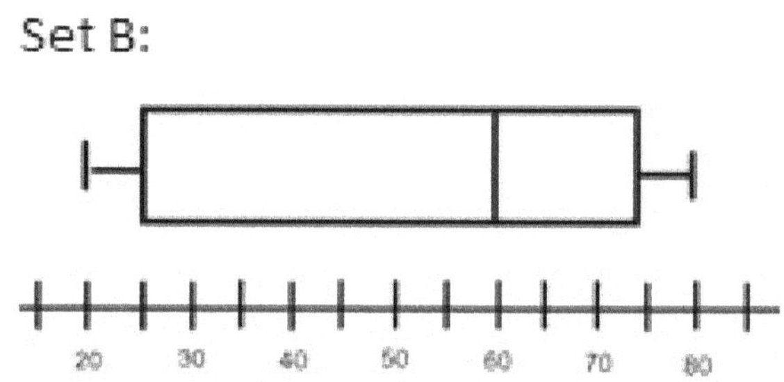

Ⓐ 30
Ⓑ 60
Ⓒ 50
Ⓓ 25

5. What is the difference between the standard deviation of Set A and the standard deviation of Set B, displayed in the dot plot below?

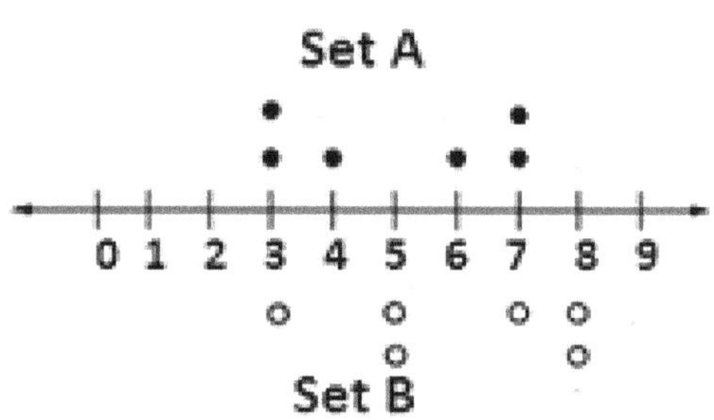

Ⓐ 0.1
Ⓑ 1.73
Ⓒ 1.83
Ⓓ 3.33

6. Which of the following are acceptable in the pre-writing stage?

Ⓐ Mental map
Ⓑ Outline
Ⓒ Free write
Ⓓ All of the above

7. How would you correct the following sentence?

The new exhibit, at the museum of modern art in new york city: is really amazing.

Ⓐ Capitalize proper nouns
Ⓑ Remove the comma after exhibit
Ⓒ Remove the colon after city
Ⓓ All of the above

8. Which of the following sentences has a dependent clause?

Ⓐ Let's go skiing tomorrow.
Ⓑ When did George get a ticket?
Ⓒ Because I love dogs, I want four when I am older.
Ⓓ I want 10 presents for Christmas and I will give 10 presents to a charity.

9. Determine whether the following statement is True or False.

Informal writing is acceptable in the drafting stage of writing.

Ⓐ True
Ⓑ False

10. When you write you should vary between complex, compound and simple sentences because this will keep your writing:

Ⓐ Challenging to others to read
Ⓑ Varied and interesting
Ⓒ Sounding smart
Ⓓ Humorous

1. Pamela is trying to determine if the after-school tutoring program helps improve student-test scores. She compares two classes test scores. None of the students in Class A participate in the program. All students in Class B participate in the after school program. The test scores for each class are below. What is the difference in the median scores of the two classes?

Class A: 60, 50, 68, 55, 65, 70, 84, 77, 78, 88, 95, 80, 40
Class B: 59, 81, 60, 68, 67, 50, 74, 92, 80, 95, 91, 78, 58

Ⓐ 70
Ⓑ 4
Ⓒ 74
Ⓓ 144

2. A data collection of the length of phone calls (in minutes) by a random sample of calls made by Stacey and Dianna is shown below. Which statement is true?

Stacey: 13, 22, 23, 24, 26, 27, 38, 39, 43, 44, 56, 57
Dianna: 13, 14, 26, 27, 28, 32, 34, 41, 42, 56, 57, 58

I - Stacey did not make any 13 minute long phone calls;
II - Stacey's median call length is lower than Dianna's median call length;
III - The mean is less than the median in both groups.

Ⓐ I Only
Ⓑ II Only
Ⓒ III Only
Ⓓ II and III Only

3. The back-to-back stem plot shows the number of sheets of paper used over several days by two teachers' students. What is the sum of the two medians of the data sets?

Mr. Smith		Ms. Jones
8	0	7
1 2 3	1	3
8	2	8
2 3 4	3	3 4
9	4	5 6 7
2	5	2 3

Ⓐ 9.5
Ⓑ 30
Ⓒ 39.5
Ⓓ 69.5

4. Two environmental science classes were studying recycling and decided to conduct surveys on how many water bottles people drank each month. The results of the surveys are shown in the back-to-back stem plot below. What is the average of the means of the survey responses? (Answers are not rounded.)

Survey I		Survey II
3	1	4
2 3 4	2	3 4
6 7	3	8
2 3	4	7 8
8	5	7 8
9	6	6 7

Ⓐ 36.7
Ⓑ 40.45
Ⓒ 44.2
Ⓓ 34.239

5. Rahim is trying to decide if regular mentally stimulating activities help delay when an elderly person develops dementia. He gathered data from two large nursing home facilities. The residents in Hifalutin Garden have regular mentally stimulating activities. The residents in Pretentious Terrace do not have regular mentally stimulating activities The age data on when patients began showing signs of dementia from his survey is below. Based on the data, what can we say about the effect of regular mental stimulation activities and the onset of dementia?

Hifalutin Garden: 83,75,70,75,78,88,102,100,92,99
Pretentious Terrace: 60,68,69,79,73,70,55,81,68,71

Ⓐ Regular mental stimulation does not appear to affect the onset of dementia
Ⓑ Regular mental stimulation appears to significantly delay the onset of dementia
Ⓒ The ages of residents in both adult facilities at which they show signs of dementia are the same
Ⓓ Regular mental stimulation appears to significantly increase the age people show signs of dementia

Effects of Video Games

Read the passage below and answer the questions that follow.

Video games have become an inescapable part of growing up and one cannot be wrong when we say that they also play a large part of development and socialization of children. Pokemon, which was very popular, has grown from just a video game into a cultural phenomenon. Technology has become more pervasive in everyday life, and with that comes the question Is technology doing good for us? The effects of video games have always been a cause of contention, right from the time they were invented. Parents have opined that video games cause children to socially isolate themselves, have been a cause of obesity, have made children insensitive to violence, and are, in short, a waste of time. However, when we look at the wide variety of games available to play, it becomes hard to decide whether all video games are bad for people in general. But there is enough evidence for both arguments; video games can be both good and bad for those that play them.

Negative effects of video games

One of the most important negative connotations that video games bring to our mind is that it increases violence and aggression in children and adolescents who play them. However, in research, there is no consensus whether video games alone have resulted in aggression or if it is more a combination of the environment, the family, or personality features. However, there is enough evidence that shows that the active participation and repeated acts of violence which comes as rewards that come from that violence in video games can result in children learning those behaviors. But then there is also evidence which says that aggressive behaviors do not carry over into real life, and when players stop playing, these violent feelings and behaviors also end.

Positive effects of video games

There are in fact several benefits of playing video games for both children as well as adolescents. Educational games during early childhood help the kids learn technology, and this is very important in a world which is increasingly technology dependent. Additionally, video games makes the learning process fun filled with the lights, colors, and interaction which helps the concept get imprinted in the child's mind. Gamification is even becoming popular in adult learning as well. Video games can even be a bonding activity between parent and child, or between friends. Video games also help children learn how to follow instructions thus inculcating discipline. Another benefit that has been seen from playing video games is the improvement in decision-making speed. Video games also improve the hand-eye coordination, fine motor skills, and spatial reasoning. The player also learns strategy and management of resources along with how to judge situations.

6. Why does the writer introduce Pokemon into the opening of her argument?

Ⓐ To emphasize positively how this video game made a culture bond.
Ⓑ To emphasize how dangerous Pokemon is for our culture.
Ⓒ To introduce the idea of how boring video games are.
Ⓓ None of the above.

7. The main purpose of this essay is to convince the reader that:

Ⓐ There are positive and negative aspects to playing video games.
Ⓑ The negatives outweigh the positive aspects of playing video games.
Ⓒ You shouldn't play video games because they aren't really educational.
Ⓓ You should only play video games because they will make you smarter.

Marriage

Read the passage below and answer the questions that follow.

Don't marry a man for money. If money is your real object, the older and uglier he is, the better; for nothing should come between you and the chosen idol of your affection. If you marry one for his money, he will find it out shortly. What sublime contempt a man must have for one who simply loves his pocket-book! Why not love his farm, or lumber-yard, or herd of cattle? The love of money is a miserly pretense of affection that leads to discontent, distrust, and disgust when they find it out. Besides, wealthy men are men of care. The wife of a noted millionaire has had her husband's body stolen from its vault, has been long kept in agony, is an object of pity to all who know her.

Another wife was heard to say, "Why, I don't have the privilege, nor the money, nor the good times that my girl Bridget enjoys. I am poor and anxious and depressed, and weary of hearing my husband say, over and over again, 'You are fixing for the poor-house.' He really thinks and believes we will end life in the poor-house; and yet he enjoys a princely income. " Thousands of such men carry their load of care, and load of wealth, and load of anxiety, and how can they carry any burden of love? Don't marry a very small man—a little fellow far below all proportion; try to get some form to admire, something to shape things to, and some one who is not lost in a crowd completely, who is too little to admire and too small for beauty.

You may need strong arms and brave hands to protect you. You will need hands to provide for and maintain you, and a good form is a fine beginning of manhood or womanhood. Mental greatness is not measured by size of brain or bodily proportions. Great men are neither always wise nor always large; they are more often of more medium build, and well balanced in gifts of mental and physical development. Of the two, a very large man is better than a small one, and a medium large woman likewise.

Don't marry too young. The right age to marry is a matter of taste; twenty-one for girls, and twenty-four for men may be a little arbitrary, but certainly is sensible. The happy early marriages are rare. It too often happens that love is mistaken, or poorly informed, or lacks an anchor in good judgment. There is no use of reasoning about it, love is love, and will marry in spite of reason, and in some cases it runs away with its choice and repents it a thousand times soon after.

But be sensible, for a life contract should be a sensible one. What is the use of throwing away one season—skipping girlhood or boyhood to rush into maturity and maternity? The records of divorce courts tell the silly and sorrowful stories of many a mismatched pair, married too young and slowly repenting of their rashness. Ask of your truest friends; take counsel; be above foolishness. Don't marry a villain. Many a girl is ripe for an adventure, and in appearance nothing more resembles an angel than a keen and designing villain—a thoroughbred; not a gambler merely, but worse, a wreck! Such men may be wary, artful, deceitful, attractive.

They are crafty; their trade compels it. They may be handsome, often so; they may be oily and slick—most of them are. They may live rich and expensive lives for a season; ill-gotten gains are not lasting. Heaven pity the girl that marries one of these adventurers, for the end is bitterness! A friend met one on the Pacific road, married him, and learned to her sorrow that he drank to excess, swore like a pirate, lived in debauchery, and early offered to swap wives for a season with a boon-companion. “And that man,” she said, “was as handsome as a dude, as slick as an auctioneer, as oily as a pedler; I loved him only one day after marriage.” Don’t marry a hypocrite. Of all things get sincerity. Get the genuine article. If you get a hypocrite, he is brass jewelry, and will easily tarnish. Make careful inquiry, see that he is all that he pretends to be, or never trust him.

The habit of deceit is one of a lifetime. Some join churches for no other reason than to cloak iniquity. It is not the rule by any means; it is a too common exception. One who goes from city to city and captivates too many by his oil of blandness; one who has no business, an idler; one who apes the rich and is ground down in poverty; one who lacks the courage to live like himself and had rather live a lie and deceive the world around him,—is an unfit companion, and will bear watching.

8. What is the structure of this text?

Ⓐ Compare/contrast
Ⓑ Problem/Solution
Ⓒ Description
Ⓓ Cause/Effect

9. What genre does this text best fit into?

Ⓐ Narrative
Ⓑ Periodical
Ⓒ Reference
Ⓓ Biography

10. Which of the following best contributes to the overall meaning?

Ⓐ Use of direct quotes
Ⓑ Use of varied sentence structure
Ⓒ Incorporating questions
Ⓓ None of the above

DAY 3

CHALLENGE YOURSELF!

✔ A Randomized Experiment to Compare Two Treatments
✔ Reading Information

www.lumoslearning.com/a/dc10-18

See the first page for Signup details

1. Suppose the function f(x) cube roots the input, doubles that value, and then adds 9. What is f(343)?

Ⓐ 25
Ⓑ 23
Ⓒ 14
Ⓓ 5

2. Suppose the function h(x) multiplies the input by 5, doubles that answer, and then adds 1. What is h(12)?

Ⓐ 25
Ⓑ 120
Ⓒ 61
Ⓓ 121

3. Suppose the function f(x) cubes twice the value of the input, then adds 3 times the input, and then subtracts 1. What is f(3)?

Ⓐ 225
Ⓑ 35
Ⓒ 13
Ⓓ 224

4. If f(x) = -3x + 5, what is the ordered pair when x = -4?

Ⓐ (17, -4)
Ⓑ (-4, 17)
Ⓒ (-4, 7)
Ⓓ (7, -4)

5. If $f(x)=\frac{2}{3}x-4$, what is the correct function notation when x=-6?

Ⓐ f(−6)=8
Ⓑ f(−6)=0
Ⓒ f(−8)=−6
Ⓓ f(−6)=−8

6. All of the following are complex sentences EXCEPT:

Ⓐ Jamal and Nicole went swimming.
Ⓑ After the game ended, we went to eat ice cream.
Ⓒ Kevin got in trouble because he failed one of his classes.
Ⓓ Even though she hates drinking milk, Jennifer can't eat cereal without it.

7. Spelling a word like GRAY as GREY represents a:

Ⓐ misspelling
Ⓑ synonym
Ⓒ antonym
Ⓓ variant

8. The word syntax is used to describe

Ⓐ the subject of a sentence
Ⓑ the way figurative language is used in a sentence
Ⓒ how vocabulary is used in a sentence
Ⓓ how sentences are put together

9. Which of the following is a complex sentence?

Ⓐ Jackson and Lauren were dating last year.
Ⓑ Jackson and Lauren haven't spoken to each other since they broke up.
Ⓒ C Jackson wanted to apologize to Lauren, but Lauren wouldn't listen.
Ⓓ Lauren wanted to forgive Jackson, so she finally answered his phone call.

10. Select the title of the source in this Works Cited entry for an article.

Coscarelli, Joe. "Taylor Swift's 'Reputation' Sells 1.2 Million Copies in Its First Week." The New York Times, 21 Nov. 2017.

Ⓐ Joe Coscarelli
Ⓑ Taylor Swift
Ⓒ Reputation
Ⓓ The New York Times

DAY 4

CHALLENGE YOURSELF!

✓ Using Function Notation & Evaluating Functions
✓ Language

www.lumoslearning.com/a/dc10-19

See the first page for Signup details

1. A ball with radius 6 cm just fits inside a cylindrical can with radius of 6 cm and a height 10 cm. How much empty space is left in the can?

Ⓐ $18\pi\ cm^3$
Ⓑ $36\pi\ cm^3$
Ⓒ $54\pi\ cm^3$
Ⓓ $72\pi\ cm^3$

2. A cone of radius r and height h is placed inside a cylindrical can of radius r and height h. What fraction of the available volume does the cone occupy?

Ⓐ $\frac{1}{9}$
Ⓑ $\frac{1}{6}$
Ⓒ $\frac{1}{3}$
Ⓓ $\frac{1}{2}$

3. A hemisphere with radius 3 cm sits on top of a cone of equal diameter and height of 10 cm as shown in the diagram below. Find the combined volume of the composite object.

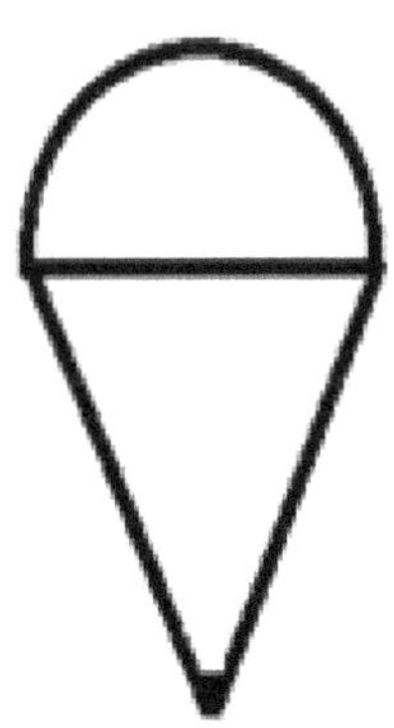

Ⓐ $24\pi cm^3$
Ⓑ $36\pi cm^3$
Ⓒ $48\pi cm^3$
Ⓓ $60\pi cm^3$

4. What is the volume of the prism shown below?

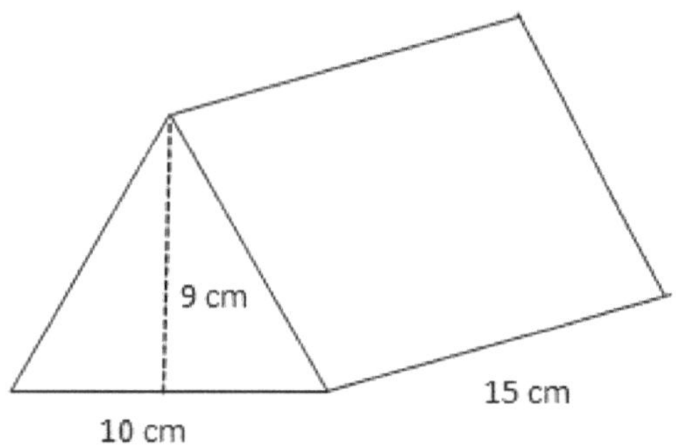

Ⓐ 1350 cm^3
Ⓑ 1350 cm
Ⓒ 675 cm^3
Ⓓ 675 cm

5. Find the volume of a cone with a diameter of 4 feet and a height of 2 feet. If necessary use π=3.14

Ⓐ 8.373 m^3
Ⓑ 8.373 m^2
Ⓒ 23.03 m^3
Ⓓ 25.12 m^3

Speech by Theodore Roosevelt

Read the passage below and answer the questions that follow.

Of course, in one sense, the first essential for a man's being a good citizen is his possession of the home virtues of which we think when we call a man by the emphatic adjective of manly. No man can be a good citizen who is not a good husband and a good father, who is not honest in his dealings with other men and women, faithful to his friends and fearless in the presence of his foes, who has not got a sound heart, a sound mind, and a sound body; exactly as no amount of attention to civil duties will save a nation if the domestic life is undermined, or there is lack of the rude military virtues which alone can assure a country's position in the world. In a free republic the ideal citizen must be one willing and able to take arms for the defense of the flag, exactly as the ideal citizen must be the father of many healthy children. A race must be strong and vigorous; it must be a race of good fighters and good breeders, else its wisdom will come to naught and its virtue be ineffective; and no sweetness and delicacy, no love for and appreciation of beauty in art or literature, no capacity for building up material prosperity can possibly atone for the lack of the great virile virtues.

But this is aside from my subject, for what I wish to talk of is the attitude of the American citizen in civic life. It ought to be axiomatic in this country that every man must devote a reasonable share of his time to doing his duty in the political life of the community. No man has a right to shirk his political duties under whatever plea of pleasure or business; and while such shirking may be pardoned in those of small cleans it is entirely unpardonable in those among whom it is most common–in the people whose circumstances give them freedom in the struggle for life. In so far as the community grows to think rightly, it will likewise grow to regard the young man of means who shirks his duty to the State in time of peace as being only one degree worse than the man who thus shirks it in time of war.

A great many of our men in business, or of our young men who are bent on enjoying life (as they have a perfect right to do if only they do not sacrifice other things to enjoyment), rather plume themselves upon being good citizens if they even vote; yet voting is the very least of their duties. Nothing worth gaining is ever gained without effort. You can no more have freedom without striving and suffering for it than you can win success as a banker or a lawyer without labor and effort, without self-denial in youth and the display of a ready and alert intelligence in middle age. The people who say that they have not time to attend to politics are simply saying that they are unfit to live in a free community.

6. What is the structure of this text?

Ⓐ Compare/contrast
Ⓑ Problem/Solution
Ⓒ Description
Ⓓ Cause/Effect

7. What genre does this text best fit into?

Ⓐ Narrative
Ⓑ Periodical
Ⓒ Reference
Ⓓ Biography

8. Which is NOT an example of a sensory detail used in a narrative writing piece?

Ⓐ I heard a rustling in the trees
Ⓑ The bitter sting of lemon grazed my tongue
Ⓒ He could not forget the locket as it fell out of his hands
Ⓓ I enjoyed the quiet ride around the lake

9. Which of the following pieces of writing are likely to be informative?

Ⓐ A set of directions to the mall.
Ⓑ An explanation of how to knit.
Ⓒ An explanation of the causes of WWI.
Ⓓ All of the above.

10. Which type of clause is represented in the underlined portion of the following sentence? "Humans and insects are similar in that they both need to breathe to survive, and both breathe out carbon dioxide."

Ⓐ independent clause
Ⓑ dependent clause
Ⓒ noun clause
Ⓓ relative clause

DAY 5

CHALLENGE YOURSELF!

✓ Finding Volume of Cylinders, Pyramids, Cones and Spheres
✓ Reading-Writing-Language

www.lumoslearning.com/a/dc10-20

See the first page for Signup details

WEEK 4 - ESSAY PROMPT

✓ An Interesting Background

www.lumoslearning.com/a/wc10-w4

See the first page for Signup details

The Ultimate Cheat Sheet for SAT/ACT Prep

1. Vocabulary

The SAT focuses a lot on your knowledge of vocabulary. The critical reading section and the essay writing is a determining factor in your final score. The more impressive your vocabulary is, the better you will score on these sections.

One way to improve your vocabulary is by doing a lot of reading. If you come across a word that you do not know or know well, look it up. Building your vocabulary will help you to break down reading passages and improve your writing.

2. Read and Write. And do it again.

The more time you spend reading and writing, the better you will become at it. Reading will build your vocabulary as you come across new words, help you see different writing styles, and introduce you to new ideas. It is a good idea to read the material you enjoy as well as some things you find boring. This will help to improve your critical thinking skills.

After you read, practice writing skills by summarizing or commenting on what you have read. This will help you to become better at expressing your ideas.

3. Use available study materials

Find some SAT and ACT prep books and put them to use. You will find test-taking tips, practice questions, reading examples, essay prompts, and more. Take advantage of the materials you can use to help you prepare and do your best.

4. Practice, Practice, Practice!

Practice tests are available to help you prepare. You will learn the style of questions asked and the kinds of prompts used for writing. This will also help you to identify any areas you may struggle with so that you can focus your studying in a way that will help you to do better.

5. Understand and memorize formulas

Sometimes you will be provided with formulas at the front of the test booklet. Though this is convenient, it can also waste time if you have to keep flipping back and forth in the exam book. If you have formulas committed to memory, it will save you time. It is also easier to apply formulas that you understand.

6. Don't cram. It creates stress.

Make sure to organize your time to study a little bit each day leading up to the test. Trying to cram the night before will create undue stress that will not have a good result.

7. Show your work.

Writing things down will help you to find a mistake if you get lost with what you were doing. It may seem silly to write out simple math problem steps, but you will be glad you did if you have to go back and check your work.

8. Drawing a blank on a math problem? Write some notes.

If you come across a problem that you are stuck on, write out the information you have. Underline important parts of the question and make diagrams to help you. Keeping your pencil moving will help to keep your mind engaged.

9. Consider the other side.

Anytime you write an essay, you need to think about the counterargument. When you acknowledge the other side of the question, your writing will become more mature and well-thought-out.

10. Analyze and Eliminate.

When you are answering multiple-choice questions, the answer may not be evident right away. You can start by eliminating the answers you know for sure are incorrect. Remember, wrong answers do not count against you, so never leave a question blank.

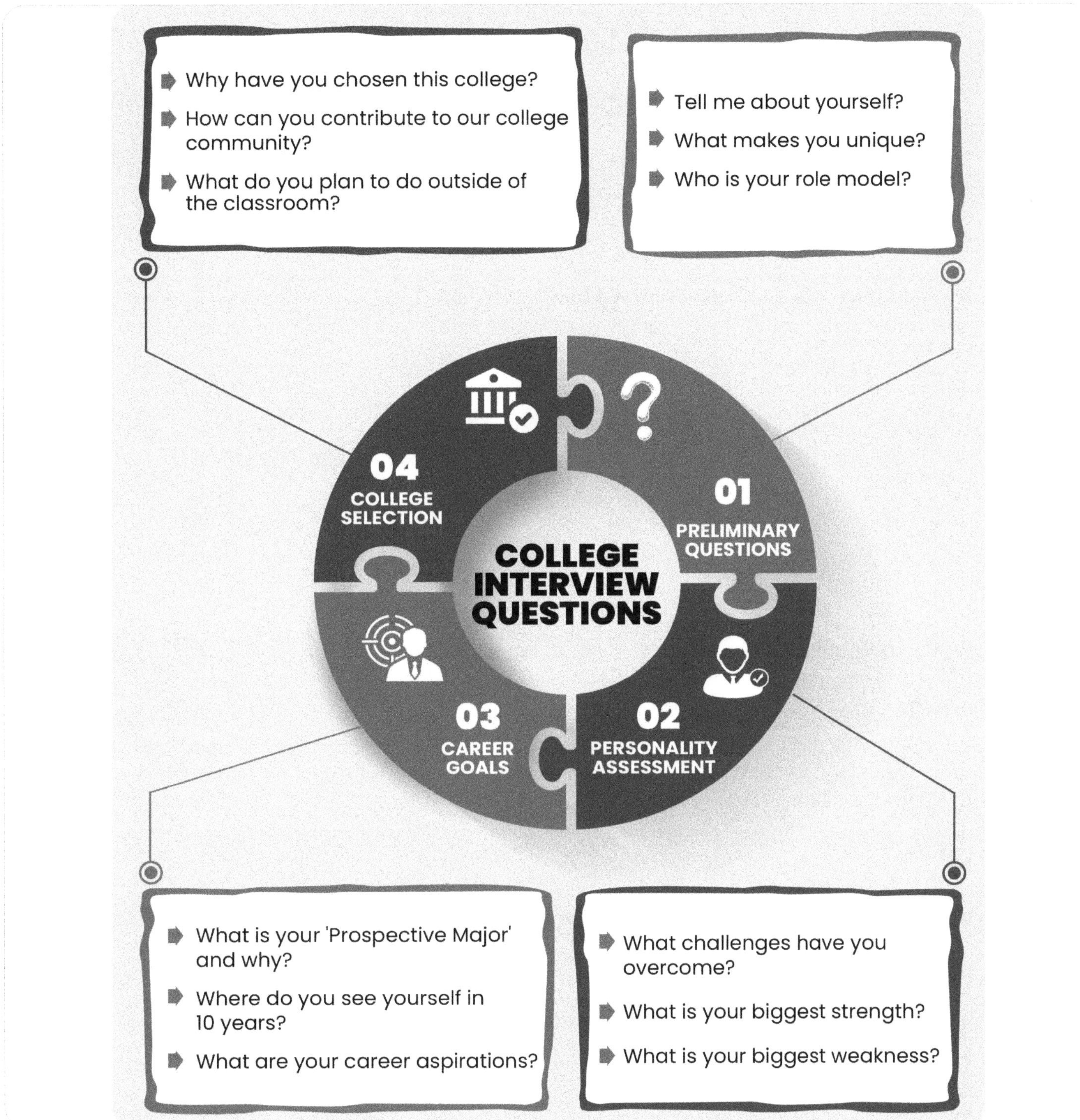

COLLEGE READINESS MIND MAPS

✓ College Interview Questions

www.lumoslearning.com/a/crm4

See the first page for Signup details

WEEK 5
SUMMER PRACTICE

SOLVE QUADRATIC EQUATIONS WITH ONE VARIABLE

DAY 1

1. What are the solutions to the quadratic equation below?

$2x^2-3x+1=0$

Ⓐ $-\frac{1}{2}, 1$

Ⓑ $\frac{1}{2}, 1$

Ⓒ $\frac{1}{2}, -1$

Ⓓ $-\frac{1}{2}, -1$

2. What are the solutions to the quadratic equation below?

$2x^2+8x+6=0$

Ⓐ -1, 3
Ⓑ 1, 3
Ⓒ 1,-3
Ⓓ -1,-3

3. What are the solutions to the quadratic equation below?

$4x^2-11x+6=0$

Ⓐ $\frac{3}{4}, 2$

Ⓑ $-\frac{3}{4}, 2$

Ⓒ $\frac{3}{4}, -2$

Ⓓ $-\frac{3}{4}, -2$

4. What are the solutions to the quadratic equation below?

$2x^2-3x+1=0$

Ⓐ $-\frac{1}{2}, 1$

Ⓑ $\frac{1}{2}, 1$

Ⓒ $\frac{1}{2}, -1$

Ⓓ $-\frac{1}{2}, -1$

5. Solve: $4x^2-8x=0$.

Ⓐ 4, −8
Ⓑ 0, −2
Ⓒ 0, 2
Ⓓ -4, 8

All the World's a Stage

Read the poem below and answer the questions that follow.

All the world's a stage,
And all the men and women merely players;
They have their exits and their entrances;
And one man in his time plays many parts,
His acts being seven ages. At first the infant,
Mewling and puking in the nurse's arms;
And then the whining school-boy, with his satchel
And shining morning face, creeping like snail
Unwillingly to school. And then the lover,
Sighing like furnace, with a woeful ballad
Made to his mistress' eyebrow. Then a soldier,
Full of strange oaths, and bearded like the pard,
Jealous in honour, sudden and quick in quarrel,
Seeking the bubble reputation
Even in the cannon's mouth. And then the justice,
In fair round belly with good capon lin'd,
With eyes severe and beard of formal cut,
Full of wise saws and modern instances;
And so he plays his part. The sixth age shifts
Into the lean and slipper'd pantaloon,
With spectacles on nose and pouch on side;
His youthful hose, well sav'd, a world too wide
For his shrunk shank; and his big manly voice,
Turning again toward childish treble, pipes
And whistles in his sound. Last scene of all,
That ends this strange eventful history,
Is second childishness and mere oblivion;
Sans teeth, sans eyes, sans taste, sans everything.

6. In the poem, the narrator addresses the 7 stages of man. Of the following, which are not stages identified:

Ⓐ Infant
Ⓑ Schoolboy
Ⓒ Lover
Ⓓ Soldier
Ⓔ Enemy

7. In the poem, the narrator describes the world as a stage and mankind has many parts. How can you infer what the narrator feels about mankind?

Ⓐ That the world is full of people who are all the same and all dependent on each other.
Ⓑ That a man's life is beautiful and demands recognition.
Ⓒ Mankind is marked by amazing stages, with each stage being a unique experience.
Ⓓ People are all different and unique.

8. In the painting, the artist captures these seven stages of man as well. Some images and stages have been brought to the foreground of the painting and stand out more. Which stage is represented by the character in the jacket carrying what looks to be a box?

Ⓐ And then the whining school-boy, with his satchel/ And shining morning face.
Ⓑ And then the justice /In fair round belly with good capon lin'd,/ With eyes severe and beard of formal cut.
Ⓒ At first the infant / Mewling and puking in the nurse's arms.
Ⓓ Is second childishness and mere oblivion.

9. In the painting, there are images that make the "world" depicted in the poem seem almost like:

Ⓐ A party
Ⓑ A circus
Ⓒ A happy street scene
Ⓓ A dog park

10. The word "reminiscence" means "a story told about a past event remembered by the narrator and it is printed on the bottom of the image. How is the poem "All the World's a Stage" similar? Which of the following is true based on what we can gather from the poem?

Ⓐ The poem and the picture are not alike because the poem is not a memory.
Ⓑ The poem is about a past event as the narrator's life closes.
Ⓒ The poem captures the stages of man almost as a memory; looking around and back at the various points in one's life.
Ⓓ All of the above.

INTERPRET THE SLOPE AND THE INTERCEPT

DAY 2

1. Alberto fills an 8-liter bucket with water to wash cars. He uses 3 liters of water to wash each car, not including rinsing, which he does with a hose. What specifically does the slope in the equation represent in this scenario?

Ⓐ The water increases by 3 liters per car
Ⓑ The water decreases by 3 liters per car
Ⓒ The water decreases by 8 liters per car
Ⓓ The water increases by 8 liters per car

2. A young tree is 8 feet tall. After 3 years, the tree is 17 feet tall. Write an equation that represents the height of the tree after any number of years. What specifically does the slope in the equation tell us about the growth of the tree?

Ⓐ The tree grows 2 feet per year
Ⓑ The tree grows 4 feet per year
Ⓒ The tree shrinks 3 feet per year
Ⓓ The tree grows 3 feet per year

3. Mike has 8 meters of a fence already completed. Each hour, he completes 4 more meters. Find how long the fence will be after Mike works for 5 more hours.

Ⓐ 12
Ⓑ 13
Ⓒ 28
Ⓓ 17

4. Given the following equation. What is the slope of the function?

$y=4x+2$

Ⓐ 4
Ⓑ 2
Ⓒ $\frac{1}{2}$
Ⓓ -4

5. What is the slope(m) and y-intercept(b) of the equation below?

$y=x$

Ⓐ m=1 b=1
Ⓑ m=0 b=1
Ⓒ m=1 b=0
Ⓓ m=0 b=0

6. Which type of music would be the best fit to accompany a presentation about homeless shelters?

Ⓐ Jazz
Ⓑ Rock
Ⓒ Ballad
Ⓓ Reggae

7. Why are the font size and type important when designing a presentation?

Ⓐ If the font and size are too big, it can distract from the presentation.
Ⓑ You want the audience to see everything presented clearly.
Ⓒ You want to be aware of your audience to choose a font and size that is appropriate for their age group.
Ⓓ All of the above.

8. How do you determine if the information you access on the internet is accurate and reliable?

Ⓐ Verify the resources by doing some research.
Ⓑ Ask someone you know for assistance.
Ⓒ If it is on the Internet, it must be true.
Ⓓ None of the above.

9. What visual media would best accompany a television show?

Ⓐ Images
Ⓑ Video
Ⓒ Data Chart
Ⓓ Comic Strip

10. If you copy information from the Internet and use it in your writing without giving credit to the source, you are:

Ⓐ Creating
Ⓑ Adjusting
Ⓒ Plagiarizing
Ⓓ None of the above

1. A drone operator flew his drone toward the ground and then at the last second turned the drone upward so the drone veered upward just above ground level. The path of the drone flight is shown below.

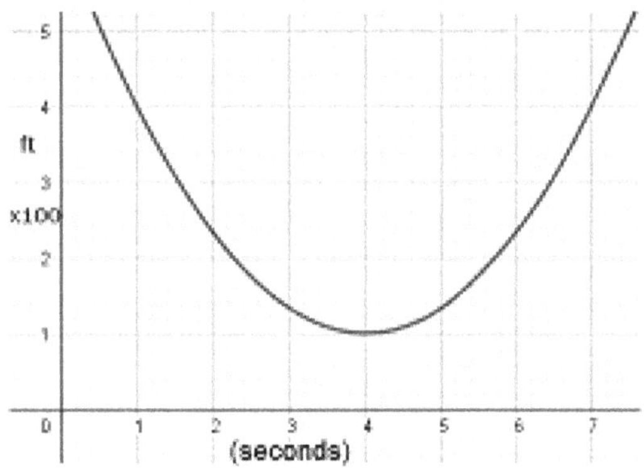

What does the lowest point on the drone's path represent?

Ⓐ The drone is at its maximum height of 100 feet above the ground at 4 seconds.
Ⓑ The drone is at its minimum height of 100 feet above the ground at 4 seconds.
Ⓒ The drone is at its maximum height of 400 feet above the ground at 1 second.
Ⓓ The drone is at its minimum height of 400 feet above the ground at 1 second.

2. An air conditioning company repairs air conditioning units 24-hours per day. The table below shows the customer complaints and compliments about repair technicians. Based on the data in the table, which shift appears to be doing the best job?

Time	Complaints	Compliments
Midnight to 6 am	3	2
6 am to Noon	4	9
Noon to 6 pm	6	12
6 pm to Midnight	8	7

Ⓐ Midnight to 6am
Ⓑ 6am to Noon
Ⓒ Noon to 6pm
Ⓓ 6pm to Midnight

3. The graph below shows the internet download speed of a company server from midnight to 8:00am. During which hour is the download speed increasing the fastest?

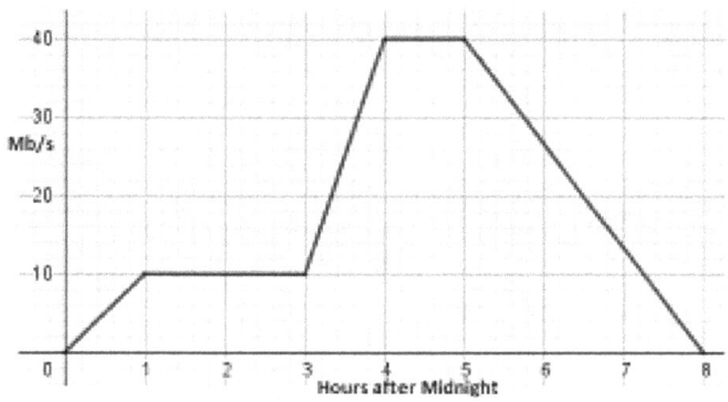

Ⓐ 4 : 00 - 5 : 00
Ⓑ 3 : 00 - 4 : 00
Ⓒ 1 : 00 - 3 : 00
Ⓓ 5 : 00 - 8 : 00

4. Which of the following is an equation of a parabola with x-intercepts at (-2, 0) and (-3, 0) and a y-intercept at (0, 6)?

Ⓐ $y=x^2-5x+6$
Ⓑ $y=x^2+5x-6$
Ⓒ $y=x^2+5x+6$
Ⓓ $y=x^2-5x-6$

5. What is the axis of symmetry for the function $y = 3x^2 + 18x + 24$?

Ⓐ $x = -3$
Ⓑ $x = 3$
Ⓒ $x = -2$
Ⓓ $x = -4$

Declaration Of Independence 1776

Read the passage below and answer the questions that follow.

We hold these truths to be self-evident: that all men are created equal; that they are endowed by their creator with inherent and* [certain] inalienable rights; that among these are life, liberty, and the pursuit of happiness: that to secure these rights, governments are instituted among men, deriving their just powers from the consent of the governed; that whenever any form of government becomes destructive of these ends, it is the right of the people to alter or abolish it, and to institute new government, laying its foundation on such principles, and organizing its powers in such form, as to them shall seem most likely to effect their safety and happiness. Prudence indeed will dictate that governments long established should not be changed for light and transient causes; and accordingly all experience hath shown that mankind are more disposed to suffer while evils are sufferable, than to right themselves by abolishing the forms to which they are accustomed. But when a long train of abuses and usurpations begun at a distinguished period and pursuing invariably the same object, evinces a design to reduce them under absolute despotism, it is their right, it is their duty to throw off such government, and to provide new guards for their future security. Such has been the patient sufferance of these colonies; and such is now the necessity which constrains them to expunge [alter] their former systems of government. The history of the present king of Great Britain is a history of unremitting [repeated] injuries and usurpations, among which appears no solitary fact to contradict the uniform tenor of the rest but all have [all having]in direct object the establishment of an absolute tyranny over these states. To prove this let facts be submitted to a candid world for the truth of which we pledge a faith yet unsullied by falsehood.

6. What argument is being made in this excerpt?

Ⓐ That all men and women are equal
Ⓑ We should be content with the rights we are given
Ⓒ That one person should not hold control over another
Ⓓ We should all live to make each other happy

7. From a rhetorical analysis of this document, which mode of persuasion is used?

Ⓐ Ethos
Ⓑ Pathos
Ⓒ Logos
Ⓓ ALL

8. Identify the tone used in the Declaration of Independence?

Ⓐ Hopeful
Ⓑ Rebellious
Ⓒ Angry
Ⓓ Content

President Theodore Roosevelt, The Roosevelt Corollary (1904)

Read the passage below and answer the questions that follow.

The excerpt below is from the Roosevelt Corollary, a speech delivered to Congress in 1904.

It is not true that the United States feels any land hunger or entertains any projects as regards the other nations of the Western Hemisphere save such as are for their welfare. All that this country desires is to see the neighboring countries stable, orderly, and prosperous. Any country whose people conduct themselves well can count upon our hearty friendship. If a nation shows that it knows how to act with reasonable efficiency and decency in social and political matters, and if it keeps order and pays its obligations, it need fear no interference from the United States.

9. In the Roosevelt Corollary, President Roosevelt's main purpose is to

Ⓐ involve itself in the domestic affairs of Latin American countries.
Ⓑ annex territories in Latin American countries for economic purposes.
Ⓒ block Latin American countries from making treaties with European powers.
Ⓓ prevent Latin American countries from buying military equipment from Europe.

10. Based on the previously shown quotations, President Roosevelt's opponents were probably most critical of his

Ⓐ weakening the military by sending troops to too many locations.
Ⓑ threatening the use of military action to dominate other nations.
Ⓒ spending money on building up economies throughout the world.
Ⓓ overstepping his constitutional authority in dealing with other nations.

1. The graph of $f(x)=\frac{1}{10}(x+)(x-2)(x-5)$ is shown below. Over which interval is f(x) positive?

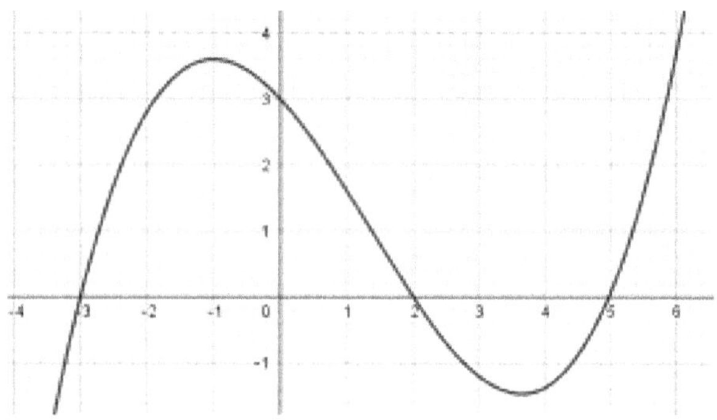

Ⓐ (5,∞)
Ⓑ (−∞,−3)U(2,5)
Ⓒ (−3,2)U(5,∞)
Ⓓ (−∞,3.43)

2. The figure below shows a polynomial function's graph. How many zeros does the function have?

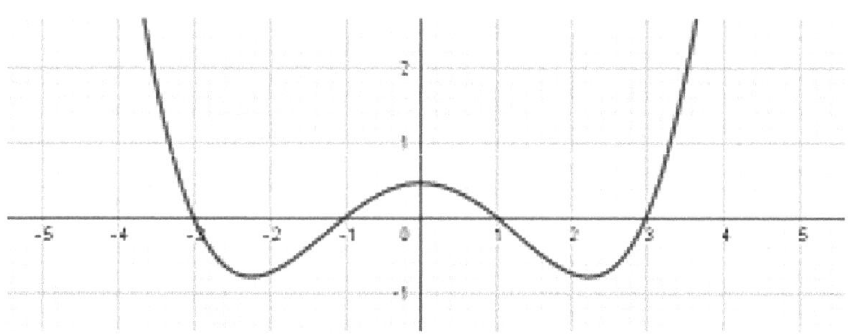

Ⓐ 3
Ⓑ 4
Ⓒ 5
Ⓓ 2

3. The figure below shows the graph of a quintic polynomial function. How many turning points does the function's graph have?

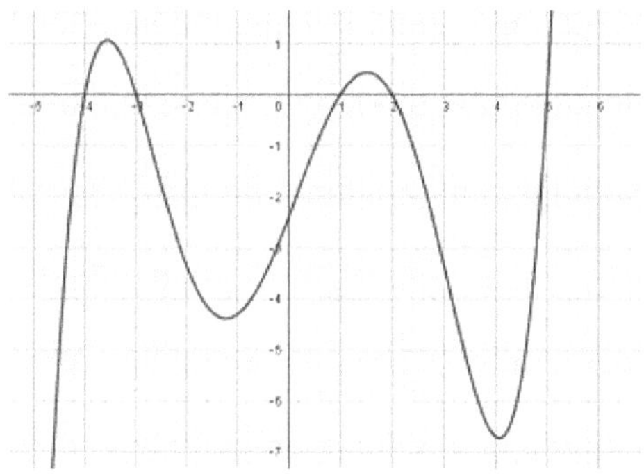

Ⓐ 3
Ⓑ 4
Ⓒ 5
Ⓓ 6

4. The figure below shows the graph of a polynomial function f(x). Is the function an odd function, an even function, or neither?

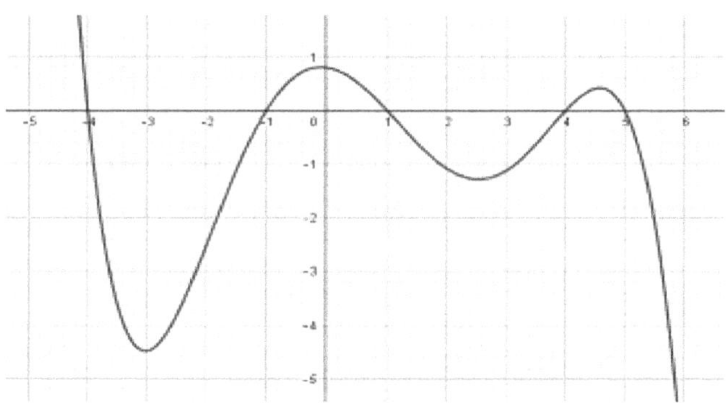

Ⓐ Neither
Ⓑ Even
Ⓒ Odd
Ⓓ Impossible to tell

5. The figure below shows the graph of a polynomial function g(x). Based on the graph, what degree is the function?

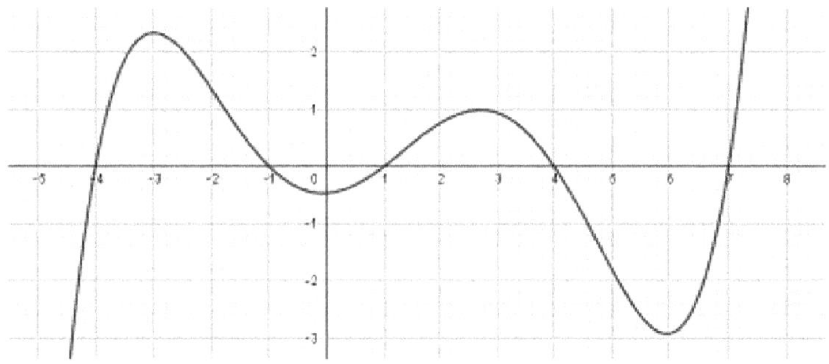

Ⓐ 6th
Ⓑ 5th
Ⓒ 4th
Ⓓ Impossible to tell

Read the following sentences.

Kevin stayed up all night. <u>Additionally</u>, he struggled to stay awake in class the next day.

6. In order to make the meaning of the two sentences more clear, which of the following would be the most precise replacement for the underlined word?

Ⓐ However
Ⓑ For example
Ⓒ Consequently
Ⓓ Indeed

7. Why should a writer consider varying his or her sentence structures in a piece of writing?

Ⓐ To take up more space.
Ⓑ Sentence structures should remain the same in order to maintain consistency.
Ⓒ Varying sentence structure avoids writing that feels monotonous and bores the reader.
Ⓓ To demonstrate knowledge of all the sentence types.

8. Which of the following punctuation marks would you use to combine two independent clauses that have closely related ideas?

Ⓐ question mark
Ⓑ comma
Ⓒ period
Ⓓ semicolon

9. Select the author's name in this Works Cited entry for a book.

Foster, Thomas C. How to read literature like a professor: A lively and entertaining guide to reading between the lines. Harper, an imprint of Harper Collins Publishers, 2017.

Ⓐ Harper Collins
Ⓑ How to Read Literature Like a Professor
Ⓒ Thomas C. Foster
Ⓓ Foster C. Thomas

10. Complete the analogy. Gasoline is to tank as money is to

Ⓐ vault
Ⓑ silo
Ⓒ mattress
Ⓓ store

1. A hopper for storing grain is built in the shape of an inverted pyramid with a square base of edge width 5 m. The full height of the pyramid is 15 m; however, the bottom 3 m was removed to provide a means to empty the hopper. Find the volume of the hopper.

Ⓐ 24 m^3
Ⓑ 54 m^3
Ⓒ 84 m^3
Ⓓ 124 m^3

2. A cone with radius 2r and height 2h as shown in the diagram below contains liquid at a level equal to half the height of the cone. What fraction of the cone's volume is occupied by the liquid?

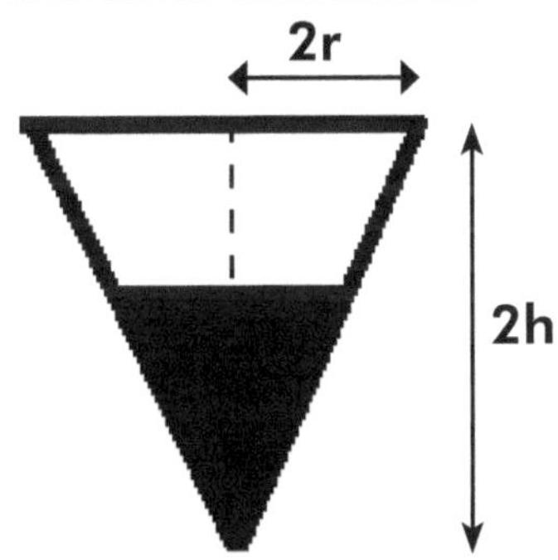

Ⓐ $\frac{1}{8}$
Ⓑ $\frac{1}{4}$
Ⓒ $\frac{3}{8}$
Ⓓ $\frac{1}{2}$

3. A cone has radius r and height h which is the same as a cylindrical can. The cone is filled with water and then poured into the can. Find the ratio of the depth of the water to the height of the can.

Ⓐ $\frac{1}{9}$
Ⓑ $\frac{1}{6}$
Ⓒ $\frac{1}{3}$
Ⓓ $\frac{1}{2}$

4. Andre wants to fill a cylindrical tank with water. If the tank has a diameter of 2 meters and is 3 meters high, how much water will Andre need? If necessary, use π=3.14.

- Ⓐ 8.14 m^3
- Ⓑ 9.14 m^3
- Ⓒ 9.42 m^3
- Ⓓ 10.25 m^3

5. Joe crafts a snowball that is 5 inches across at the diameter. Assume the snowball is a perfect sphere. Which value best represents the volume of the snowball? Use π=3.14. Round to the nearest whole number.

- Ⓐ 47 in^3
- Ⓑ 65 in^3
- Ⓒ 52 in^3
- Ⓓ 57 in^3

NYC

Read the passage below and answer the question that follows.

NYC is one of the most amazing cities in the world and Manhattan at only 22.82 miles has more to offer people than any place in the world. You can start your exploration down at the southern tip. Walk through Wall Street and stop to admire the Wall Street bull in all of its bronze glory. Continue walking and each neighborhood from Chinatown to Little Italy to SoHo and the West Village offer fresh smells and sites. The shopping and eating in this part of the city are unlike any other city. Keep walking and pass the fashion district and the street of shoes! Every neighborhood drips in culture and if you want to visit the art museums, check out museum mile on 5th. There's a museum for you! Be sure to visit soon.

6. Choose the best summary of this passage.

Ⓐ The neighborhoods of Manhattan are varied and there is something for everyone.
Ⓑ There are many art museums in Manhattan.
Ⓒ NYC and Manhattan are the same thing.
Ⓓ The food is the best in the world.

Quakers

Read the passage below and answer the question that follows.

Quakers . . . forbid any act that acknowledged the right of slavery. Quaker meetings in western New York advised [Quakers] not to use any product ”no cotton cloth, no white sugar” made with the unpaid labor of slaves. Good Quakers used maple sugar in their coffee. - Christopher Densmore, Historian

7. Based on the provided text, Quakers reacted to slavery by

Ⓐ publishing pro-slavery newspapers.
Ⓑ declining to sell items from slave states.
Ⓒ speaking publicly about the horrors of slavery.
Ⓓ refusing to purchase goods produced by slaves.

8. In the below plot chart, which event is C:

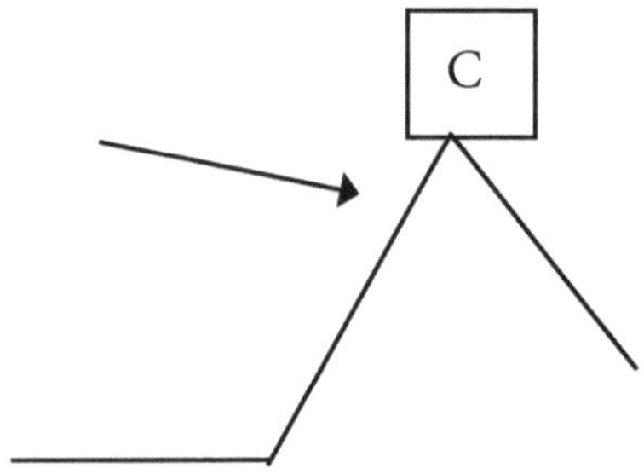

Ⓐ Rising Action
Ⓑ Climax
Ⓒ Exposition
Ⓓ Falling Action

9. You are absolutely disgusted with how your town handles waste, and you want something done about it. Which type of writing would you use?

Ⓐ Narrative
Ⓑ Exposition
Ⓒ Persuasive
Ⓓ Informative

10. What type of writing is this?

Costa Rica is a wonderful place to visit, and you should come here too. The amazing green and blue of the country invite you in to really feel enveloped in the luxury of the country.

Ⓐ Narrative
Ⓑ Exposition
Ⓒ Persuasive
Ⓓ Informative

Reading Strategies and Tips to Improve SAT, ACT Score

Reading fluency will help you improve your SAT and ACT scores. The first important step is to find books that you enjoy. Reading can be boring if you do not have something that engages you. Once you find the kind of books you enjoy, read as much as you can.

SAT and ACT

The more you read, the more your reading fluency will improve. The verbal section of the SAT is based on reading and understanding passages. Since it is a timed test, you will need to read at a certain pace to get all of the questions answered. Even if you have a high level of understanding, that does not help much if you read at a slow pace. The more you read every day, the faster you will be able to read and comprehend the passages on the test. When you are a more efficient reader, you will have more time to answer the questions on the test. The ACT measures college readiness, and to be accepted into a college, you need an acceptable level of reading fluency as well.

Strategies Used to Evaluate Reading Fluency

To assess fluency, evaluators often use timed oral reading tests and then ask comprehension questions at the end of the reading. On a standardized evaluation, there will be reading comprehension and vocabulary sections that are used to assess the applicant's fluency.

Improve Your Reading Fluency

The best thing you can do to improve your fluency is to read every day. The more you read, the more fluent you become. Read to yourself and then put the book down and see if you can summarize what you have read.

Tips for the Test

1. Identify the mistakes you make most often when you are reading so you can focus on improving in that area.
2. Practice skimming passages.
3. Skip difficult questions and return to them later.
4. Use the process of elimination when needed.
5. Practice!

The SAT

On the SAT, the first section is reading comprehension and vocabulary. You will be provided five passages that you will need to read and then you will have 52 multiple choice questions to answer based on the reading. This section of the test will evaluate your command of evidence and words in context. To get ready, you should practice reading passages similar to what will be found on the SAT.

The ACT

The American College Test is like the SAT and will test the same skills. To do well on this test, you will need to be able to read fluently. Reading fluently will help to improve both comprehension and vocabulary.

Extra Tips

Though it can be more difficult to improve fluency at an older age, it is an attainable goal. Read as much as you can and read out loud when you can. It is also a good idea to listen to other readers and pay attention to their pace. Audiobooks are a good way to do this. Practice will improve your skills.

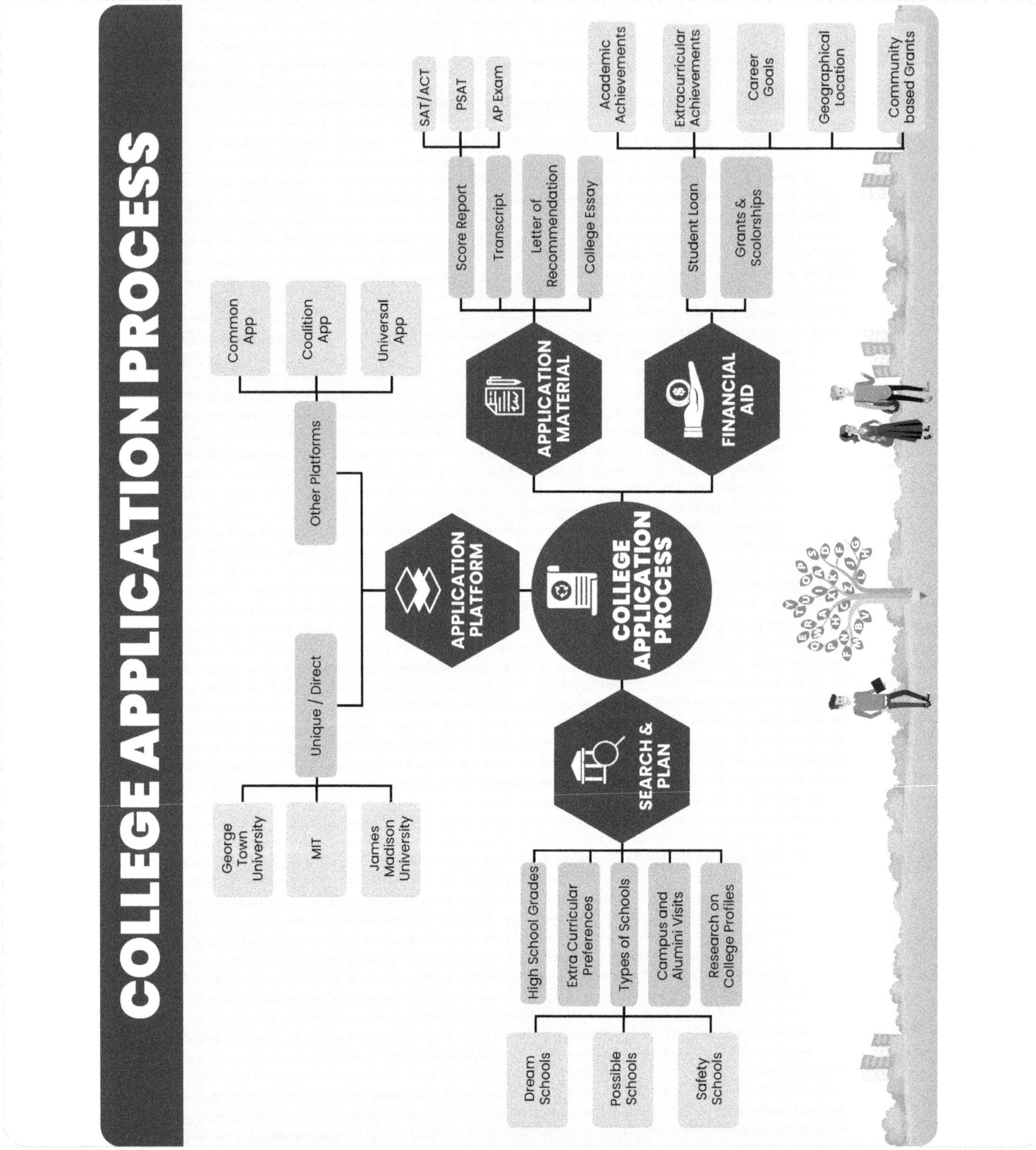
COLLEGE APPLICATION PROCESS
COLLEGE APPLICATION PROCESS
SEARCH & PLAN
High School Grades
Extra Curricular Preferences
Types of Schools
Campus and Alumini Visits
Research on College Profiles
Dream Schools
Possible Schools
Safety Schools
APPLICATION PLATFORM
Unique / Direct
George Town University
MIT
James Madison University
Other Platforms
Common App
Coalition App
Universal App
APPLICATION MATERIAL
Score Report
Transcript
Letter of Recommendation
College Essay
SAT/ACT
PSAT
AP Exam
FINANCIAL AID
Student Loan
Grants & Scolorships
Academic Achievements
Extracurricular Achievements
Career Goals
Geographical Location
Community based Grants

WEEK 6
SUMMER PRACTICE

SOLVE QUADRATIC EQUATIONS WITH ONE VARIABLE

DAY 1

1. What are the solution values of x for the quadratic equation below?

$$4x^2+8x-5=0$$

Ⓐ $-\frac{1}{2}, \frac{5}{2}$
Ⓑ $\frac{1}{2}, -\frac{5}{2}$
Ⓒ $-\frac{1}{2}, -\frac{5}{2}$
Ⓓ $\frac{1}{2}, \frac{5}{2}$

2. What are the solution values of x for the quadratic equation below?

$$4x^2+8x+4=0$$

Ⓐ 1
Ⓑ -2,1
Ⓒ -1
Ⓓ -2,-1

3. What are the solution values of x for the quadratic equation below?

$$4x^2-12x+9=0$$

Ⓐ $\frac{-3}{2}, -1$
Ⓑ $\frac{-3}{2}, 1$
Ⓒ $\frac{3}{2}, \frac{-3}{2}$
Ⓓ $\frac{3}{2}, \frac{3}{2}$

4. What are the solution values of x for the quadratic equation below?

$x^2+19x+60=0$

Ⓐ -15, -4
Ⓑ -15, 4
Ⓒ 15, -4
Ⓓ 15, 4

5. Solve the quadratic $x^2+10x=-25$.

Ⓐ -10
Ⓑ 10
Ⓒ 5
Ⓓ -5

Down the Rabbit Hole

Read the passage below and answer the questions that follow.

(1) Alice was beginning to get very tired of sitting by her sister on the bank, and of having nothing to do: once or twice she had peeped into the book her sister was reading, but it had no pictures or conversations in it, 'and what is the use of a book,' thought Alice 'without pictures or conversations?'

(2) So she was considering in her own mind (as well as she could, for the hot day made her feel very sleepy and stupid), whether the pleasure of making a daisy-chain would be worth the trouble of getting up and picking the daisies, when suddenly a White Rabbit with pink eyes ran close by her.

(3) There was nothing so very remarkable in that; nor did Alice think it so very much out of the way to hear the Rabbit say to itself, 'Oh dear! Oh dear! I shall be late!' (when she thought it over afterwards, it occurred to her that she ought to have wondered at this, but at the time it all seemed quite natural); but when the Rabbit actually took a watch out of its waistcoat-pocket, and looked at it, and then hurried on, Alice started to her feet, for it flashed across her mind that she had never before seen a rabbit with either a waistcoat-pocket, or a watch to take out of it, and burning with curiosity, she ran across the field after it, and fortunately was just in time to see it pop down a large rabbit-hole under the hedge.

(4) In another moment down went Alice after it, never once considering how in the world she was to get out again.

(5) The rabbit-hole went straight on like a tunnel for some way, and then dipped suddenly down, so suddenly that Alice had not a moment to think about stopping herself before she found herself falling down a very deep well.

(6) Either the well was very deep, or she fell very slowly, for she had plenty of time as she went down to look about her and to wonder what was going to happen next. First, she tried to look down and make out what she was coming to, but it was too dark to see anything; then she looked at the sides of the well, and noticed that they were filled with cupboards and book-shelves; here and there she saw maps and pictures hung upon pegs. She took down a jar from one of the shelves as she passed; it was labeled 'ORANGE MARMALADE', but to her great disappointment it was empty: she did not like to drop the jar for fear of killing somebody, so managed to put it into one of the cupboards as she fell past it.

(7) 'Well!' thought Alice to herself, 'after such a fall as this, I shall think nothing of tumbling down stairs! How brave they'll all think me at home! Why, I wouldn't say anything about it, even if I fell off the top of the house!' (Which was very likely true.)

(8) Down, down, down. Would the fall never come to an end! 'I wonder how many miles I've fallen by this time?' she said aloud. 'I must be getting somewhere near the center of the earth. Let me see: that would be four thousand miles down, I think—' (for, you see, Alice had learned several things of this sort in her lessons in the schoolroom, and though this was not a very good opportunity for showing off her knowledge, as there was no one to listen to her, still it was good practice

to say it over) '—yes, that's about the right distance—but then I wonder what Latitude or Longitude I've got to?' (Alice had no idea what Latitude was, or Longitude either, but thought they were nice grand words to say.)

(9) Presently she began again. 'I wonder if I shall fall right through the earth! How funny it'll seem to come out among the people that walk with their heads downward! The Antipathies, I think—' (she was rather glad there was no one listening, this time, as it didn't sound at all the right word) '—but I shall have to ask them what the name of the country is, you know. Please, Ma'am, is this New Zealand or Australia?' (and she tried to curtsy as she spoke—fancy curtsying as you're falling through the air! Do you think you could manage it?) 'And what an ignorant little girl she'll think me for asking! No, it'll never do to ask: perhaps I shall see it written up somewhere.'

(10) Down, down, down. There was nothing else to do, so Alice soon began talking again. 'Dinah'll miss me very much to-night, I should think!' (Dinah was the cat.) 'I hope they'll remember her saucer of milk at tea-time. Dinah, my dear! I wish you were down here with me! There are no mice in the air, I'm afraid, but you might catch a bat, and that's very like a mouse, you know. But do cats eat bats, I wonder?' And here Alice began to get rather sleepy, and went on saying to herself, in a dreamy sort of way, 'Do cats eat bats? Do cats eat bats?' and sometimes, 'Do bats eat cats?' for, you see, as she couldn't answer either question; it didn't much matter which way she put it. She felt that she was dozing off, and had just begun to dream that she was walking hand in hand with Dinah, and saying to her very earnestly, 'Now, Dinah, tell me the truth: did you ever eat a bat?' when suddenly, thump! thump! down she came upon a heap of sticks and dry leaves, and the fall was over.

(11)Alice was not a bit hurt, and she jumped up on to her feet in a moment: she looked up, but it was all dark overhead; before her was another long passage, and the White Rabbit was still in sight, hurrying down it. There was not a moment to be lost: away went Alice like the wind, and was just in time to hear it say, as it turned a corner, 'Oh my ears and whiskers, how late it's getting!' She was close behind it when she turned the corner, but the Rabbit was no longer to be seen: she found herself in a long, low hall, which was lit up by a row of lamps hanging from the roof.

(12) There were doors all round the hall, but they were all locked; and when Alice had been all the way down one side and up the other, trying every door, she walked sadly down the middle, wondering how she was ever to get out again.

(13) Suddenly she came upon a little three-legged table, all made of solid glass; there was nothing on it except a tiny golden key, and Alice's first thought was that it might belong to one of the doors of the hall; but, alas! Either the locks were too large, or the key was too small, but at any rate, it would not open any of them. However, on the second time round, she came upon a low curtain she had not noticed before, and behind it was a little door about fifteen inches high: she tried the little golden key in the lock, and to her great delight it fitted!

(14) Alice opened the door and found that it led into a small passage, not much larger than a rat-hole: she knelt and looked along the passage into the loveliest garden you ever saw. How she longed to get out of that dark hall and wander about among those beds of bright flowers and those cool fountains, but she could not even get her head through the doorway; 'and even if my head would go through,' thought poor Alice, 'it would be of very little use without my shoulders. Oh, how I wish I could shut up like a telescope! I think I could if I only knew how to begin.' For, you see, so many out-of-the-way things had

happened lately, that Alice had begun to think that very few things indeed were impossible.

(15) There seemed to be no use in waiting by the little door, so she went back to the table, half hoping she might find another key on it, or at any rate a book of rules for shutting people up like telescopes: this time she found a little bottle on it, ('which certainly was not here before,' said Alice,) and round the neck of the bottle was a paper label, with the words 'DRINK ME' beautifully printed on it in large letters.

(16) It was all very well to say 'Drink me,' but the wise little Alice was not going to do that in a hurry. 'No, I'll look first,' she said, 'and see whether it's marked "poison" or not'; for she had read several nice little histories about children who had got burnt, and eaten up by wild beasts and other unpleasant things, all because they would not remember the simple rules their friends had taught them: such as, that a red-hot poker will burn you if you hold it too long; and that if you cut your finger very deeply with a knife, it usually bleeds; and she had never forgotten that, if you drink much from a bottle marked 'poison,' it is almost certain to disagree with you, sooner or later.

(17) However, this bottle was not marked 'poison,' so Alice ventured to taste it, and finding it very nice, (it had, in fact, a sort of mixed flavor of cherry-tart, custard, pine-apple, roast turkey, toffee, and hot buttered toast,) she very soon finished it off.

6. Alice's character is established in the opening paragraph. What can we infer about Alice's character based on the comment:

"and what is the use of a book,' thought Alice 'without pictures or conversations?'

Ⓐ She is a boring person.
Ⓑ She doesn't like her sister.
Ⓒ She is silly.
Ⓓ She is curious about people and their conversations.

7. In paragraphs 5-7, what tone is established by the writer as Alice moves through the tunnel?

Ⓐ Scared
Ⓑ Curious
Ⓒ Awestruck
Ⓓ Assertive

In paragraph 9, Carroll chooses to reveal Alice's lack of knowledge because, "The Antipathies, I think--' (she was rather glad there was no one listening, this time, as it didn't sound at all the right word)."

8. Why do you think the author chooses to reveal this lack of knowledge about Alice? How does it add to her character?

Ⓐ It makes Alice seem silly, so the reader feels sorry for her.
Ⓑ It makes Alice seem more human so that the reader connects with her.
Ⓒ It encourages the reader to learn a new word.
Ⓓ It encourages the reader to think of the book.

9. When Alice lands at the bottom, she tries to get outside. She realizes she can't and says Oh, how I wish I could shut up like a telescope!" In this sentence, the author uses which literary device?

Ⓐ Personification
Ⓑ Simile
Ⓒ Metaphor
Ⓓ Theme

10. When Alice says "Oh how I wish I could shut up like a telescope!", what trait of a telescope does she wish she had?

Ⓐ Telescopes can see things from far away.
Ⓑ Telescopes can be very useful tools.
Ⓒ Telescopes have a lens that shrinks to see things from far away.
Ⓓ Telescopes can contract and expand in size

INTRODUCTION TO IMAGINARY AND COMPLEX NUMBERS

DAY 2

1. Given that $i^2 = -1$, What is $\sqrt{144} + \sqrt{-121}$?

- Ⓐ 12+11i
- Ⓑ 12-11i
- Ⓒ 11+12i
- Ⓓ 11-12i

2. Given that $i^2 = -1$, what is $\sqrt{25} - \sqrt{-36}$?

- Ⓐ 30
- Ⓑ -30
- Ⓒ 30i
- Ⓓ -30i

3. Given that $i^2 = -1$ what is i^{-451}?

- Ⓐ i
- Ⓑ -1
- Ⓒ -i
- Ⓓ 1

4. Given that $i^2=-1$, what is $\sqrt{-256} + \sqrt{225}$?

- Ⓐ 15+16i
- Ⓑ 31
- Ⓒ 15-16i
- Ⓓ -1

5. What does 3 + 7i simplify to be?

- Ⓐ 10i
- Ⓑ 21i
- Ⓒ $10i^2$
- Ⓓ 3 + 7i

WRITING

DAY 2

6. What is it called when you let the reader know where your source came from?

Ⓐ Citing Sources
Ⓑ Repeating sources
Ⓒ Doctoring sources
Ⓓ Bibliographing sources

7. Which of the following should be characteristics of a good research topic?

Ⓐ focused on a few variables
Ⓑ can be researched and proven with evidence
Ⓒ the topic should be substantial
Ⓓ all of the above

8. Which of the following are styles you can use to cite and reference your paper?

Ⓐ MLA
Ⓑ APA
Ⓒ Chicago
Ⓓ All of the above

9. Where would you place a citation like this in your research paper:

Havard, Frances. "Teacher on a Mission: Saving Students from Plagiairism." New York Times, 22 May 2017, www.francestimes.com/2017/05/22/teachers.html?_r=0. Accessed 17 May 2018.

Ⓐ Works Cited page
Ⓑ Introduction
Ⓒ Right after you use a quote from the article in the body of your essay
Ⓓ In the conclusion

10. How should you offset a direct quote that is less than four lines from a source in your writing?

Ⓐ Use a colon to indicate a quote is beginning
Ⓑ Put the quote within quotation marks
Ⓒ Put the quote in italics
Ⓓ Underline the quote

DAY 2

CHALLENGE YOURSELF!

✓ Introduction to Imaginary and Complex Numbers
✓ Writing

www.lumoslearning.com/a/dc10-27

See the first page for Signup details

ADD, SUBTRACT, AND MULTIPLY COMPLEX NUMBERS

DAY 3

1. Given that $i^2 = -1$, what is $\sqrt{(-25)} \cdot \sqrt{(-36)}$?

Ⓐ 30
Ⓑ -30
Ⓒ 30i
Ⓓ -30i

2. Given that $i^2 = -1$ what is $\sqrt{(-256)} + \sqrt{(-225)}$?

Ⓐ 31i
Ⓑ 31
Ⓒ -31
Ⓓ -31i

3. Given that $i^2 = -1$, what is $(2+3i)\times(4+i)$?

Ⓐ 11+14i
Ⓑ 11-14i
Ⓒ 5+14i
Ⓓ 5-14i

4. Given that $i^2 = -1$, what is $(-5+3i)\times(-4-9i)$?

Ⓐ 47-33i
Ⓑ -47+33i
Ⓒ -47-33i
Ⓓ 47+33i

5. Simplify: $(6i)^2$?

Ⓐ -36
Ⓑ $36i^2$
Ⓒ -12
Ⓓ $12i^2$

Ending the Stigma Around Mental Health Problems: A Path to Healing

Read the passage below and answer the questions that follow.

Mental health problems affect millions of people worldwide, yet stigma surrounding these issues remains a significant barrier to treatment and understanding. This stigma can manifest in many ways, including discrimination, judgment, and the perception that mental health issues are a sign of weakness. As a result, many individuals suffering from mental health conditions may avoid seeking help, leading to untreated symptoms and worsened conditions. However, there are several effective solutions to help reduce and eventually end the stigma around mental health.

One key solution to combating mental health stigma is education and awareness. Many people harbor negative attitudes toward mental health simply because they lack understanding. By educating the public about mental health conditions, their causes, and treatments, society can replace misconceptions with facts. Schools, workplaces, and media outlets can all play a role in disseminating accurate information. Public campaigns and mental health advocacy programs, such as Mental Health Awareness Month, can help raise visibility and normalize discussions about mental health, making it as accepted and understood as physical health.

Another important step is to encourage open conversations. Often, people hesitate to discuss mental health issues because they fear being judged or misunderstood. Creating safe spaces for dialogue, whether in workplaces, schools, or social settings, can help people feel supported and validated. Encouraging celebrities, public figures, and community leaders to share their personal mental health struggles can also help normalize the experience. The more people talk openly about their mental health, the less likely it is that these issues will remain shrouded in secrecy and shame.

Reducing discrimination in workplaces, schools, and healthcare settings is also critical. Policies that protect individuals from discrimination based on mental health status can foster an environment where people feel safe seeking help. Providing mental health resources and support in these settings, such as access to counseling or employee assistance programs, can ensure that those struggling with mental health issues receive the support they need.

Finally, supporting mental health treatment is essential in reducing stigma. Ensuring that therapy, counseling, and psychiatric services are easily accessible and affordable can encourage people to seek professional help without fear of judgment. In addition, integrating mental health care into primary healthcare settings can help make mental health services a regular part of healthcare, further normalizing their importance.

In conclusion, ending the stigma surrounding mental health requires a combination of education, open conversations, anti-discrimination policies, and accessible treatment options. By working together to address these issues, we can create a society where individuals feel empowered to seek help and receive the support they need to live healthy, fulfilling lives.

6. What is the main focus of the passage?

Ⓐ The causes of mental health issues
Ⓑ Ways to end the stigma surrounding mental health problems
Ⓒ The impact of mental health problems on physical health
Ⓓ The history of mental health treatments

7. What is one solution suggested in the passage to reduce mental health stigma?

Ⓐ Ignoring mental health issues
Ⓑ Promoting more physical health awareness
Ⓒ Educating the public about mental health conditions
Ⓓ Reducing the number of mental health professionals

8. How can public campaigns contribute to ending mental health stigma, according to the passage?

Ⓐ By providing resources for treatment only
Ⓑ By spreading awareness and normalizing conversations about mental health
Ⓒ By increasing restrictions on discussing mental health in public
Ⓓ By offering financial support for mental health patients

9. What role do celebrities and public figures play in reducing mental health stigma?

Ⓐ They discourage people from talking about mental health
Ⓑ They promote physical health over mental health
Ⓒ They share their own mental health struggles to normalize the conversation
Ⓓ They create policies to regulate mental health care

Hamilton

Read the passage below and answer the questions that follow.

To the People of the State of New York: AFTER an unequivocal experience of the inefficacy of the subsisting federal government, you are called upon to deliberate on a new Constitution for the United States of America. The subject speaks its own importance; comprehending in its consequences nothing less than the existence of the UNION, the safety and welfare of the parts of which it is composed, the fate of an empire in many respects the most interesting in the world. It has been frequently remarked that it seems to have been reserved to the people of this country, by their conduct and example, to decide the important question, whether societies of men are really capable or not of establishing good government from reflection and choice, or whether they are forever destined to depend for their political constitutions on accident and force. If there be any truth in the remark, the crisis at which we are arrived may with propriety be regarded as the era in which that decision is to be made; and a wrong election of the part we shall act may, in this view, deserve to be considered as the general misfortune of mankind.

This idea will add the inducements of philanthropy to those of patriotism, to heighten the solicitude which all considerate and good men must feel for the event. Happy will it be if our choice should be directed by a judicious estimate of our true interests, unperplexed and unbiased by considerations not connected with the public good. But this is a thing more ardently to be wished than seriously to be expected. The plan offered to our deliberations affects too many particular interests, innovates upon too many local institutions, not to involve in its discussion a variety of objects foreign to its merits, and of views, passions and prejudices little favorable to the discovery of truth.

10. The Bill of Rights guarantees certain rights to all persons accused of crimes. These rights include a trial by jury and the right to be represented by a lawyer. Why is it important to protect the rights of the accused?

Ⓐ to ensure that trials are fair
Ⓑ to increase the power of judges
Ⓒ to prevent criminals from escaping
Ⓓ to encourage people to serve on juries

MODELING LINEAR AND EXPONENTIAL FUNCTIONS

1. One dollar is invested in an account that accumulates 20% interest every year. What type of function would represent this situation?

Ⓐ linear
Ⓑ quadratic
Ⓒ exponential growth
Ⓓ exponential decay

2. The pressure in a chamber of a compressed oxygen tank begins at -2 PSI (pounds per square inch) and every day, the pressure increases 20% more than the previous day. What type of function would represent this situation?

Ⓐ linear
Ⓑ quadratic
Ⓒ exponential growth
Ⓓ exponential decay

3. Nicholas invested $200 in a long-term investment account that accumulates a guaranteed 1% every six months, indefinitely. What type of function would represent this situation?

Ⓐ linear
Ⓑ quadratic
Ⓒ exponential growth
Ⓓ exponential decay

4. An elderly man has $40,000 in a special deferred tax retirement account. Because of tax rules, he has to withdraw the same specific amount every year so that the account is empty in 10 years. and the taxes on the account are paid in full. What type of function would represent this situation?

Ⓐ linear
Ⓑ quadratic
Ⓒ exponential growth
Ⓓ exponential decay

5. A dog food plant produces 20 tons of dog food each month. Let y be the total tons of dog food produced and x be the number of months. Which of the following is the best function to model this situation?

Ⓐ $y=20^x+20x$
Ⓑ $y=20^x$
Ⓒ $y=20x$
Ⓓ None of these

6. Use the context of the surrounding sentence to determine the meaning of the underlined word.

While most people achieve success when they're older, Miles was a wunderkind, having created his own software company at age sixteen.

Ⓐ someone extraordinary
Ⓑ normal person
Ⓒ exception
Ⓓ person who achieves success at a young age

7. Read the sentence. Based on the context, determine the meaning of the underlined word.

"It was the Student Council president's job to disseminate information about the school events to the students in her class, making sure everyone was informed."

Ⓐ spread
Ⓑ keep secret
Ⓒ disorganize
Ⓓ decide

8. Read the sentence. Based on the context, determine the meaning of the underlined word.

"In countries where the government is controlled by a dictator, officials usually jail dissidents who disagree with the laws."

Ⓐ criminals
Ⓑ protesters
Ⓒ teachers
Ⓓ government officials

9. Use the context of the surrounding sentence to determine the meaning of the underlined word.

Because no event like it had ever happened before, the Astros' World Series win was unprecedented.

Ⓐ unheard of
Ⓑ expected
Ⓒ significant
Ⓓ unimportant

10. **Read the sentence. Based on the context, determine the meaning of the underlined word.**

"The teacher kept the class running smoothly by facilitating the lesson sequence, keeping everyone on task."

Ⓐ controlling
Ⓑ reporting
Ⓒ enabling
Ⓓ cancelling

EQUATION OF A CIRCLE

1. Find the equation of the circle that has center at (3, -4) and radius = 11.

Ⓐ $(x-3)^2+(y+4)^2=121$
Ⓑ $(x-3)^2+(y-4)^2=121$
Ⓒ $(x+3)^2+(y-4)^2=121$
Ⓓ $(x-3)^2+(y+4)^2=11$

2. Find the correct transformation of the circle originally based at the origin, whose equation is $(x-2)^2+(y+3)^2=16$.

Ⓐ left two, up 3; r = 4
Ⓑ left two, up 3; r = 16
Ⓒ right two, down 3, r = 16
Ⓓ right two, down 3; r = 4

3. If a circle has an equation of $(x-5)^2+(y+3)^2=36$, what are the coordinates of the center and the value of the diameter?

Ⓐ C(5, -3); d = 12
Ⓑ C(-5, 3); d = 12
Ⓒ C(5, -3); d = 6
Ⓓ C(-5, 3); d = 6

4. What is the equation of a circle, in the form, $(x-h)^2+(y-k)^2=r^2$, if the center of the circle is at the point (1,−3) and the point (4,0) is on the circle?

Ⓐ $(x-1)^2+(y+3)^2=3\sqrt{2}$
Ⓑ $(x-1)^2+(y+3)^2=18$
Ⓒ $(x+1)^2+(y-3)^2=18$
Ⓓ $(x+1)^2+(y-3)^2=3\sqrt{2}$

5. What is the equation of a circle, in the form, $(x-h)^2+(y-k)^2=r^2$, if the center of the circle is at the point (1,−4) and the point (6,4) is on the circle?

Ⓐ $(x-1)^2+(y+4)^2=\sqrt{89}$
Ⓑ $(x+1)^2+(y-4)^2=\sqrt{89}$
Ⓒ $(x-1)^2+(y+4)^2=89$
Ⓓ $(x+1)^2+(y-4)^2=89$

Preamble

Read the passage below and answer the questions that follow.

We the People of the United States, in Order to form a more perfect Union, establish Justice, insure domestic Tranquility, provide for the common defense, promote the general Welfare, and secure the Blessings of Liberty to ourselves and our Posterity, do ordain and establish this Constitution for the United States of America.

6. What is the historical significance of this work?

Ⓐ Equality amongst all men
Ⓑ To justify the need for the constitution
Ⓒ The first establishment of rules and laws in the United States
Ⓓ Create order among the people

7. The main purpose of this preamble is to

Ⓐ persuade
Ⓑ inform
Ⓒ describe
Ⓓ entertain

Merry Autumn

by Paul Laurence Dunbar

Read the poem below and answer the questions that follow.

Now purple tints are all around;
The sky is blue and mellow;
And e'en the grasses turn the ground
From modest green to yellow...

A butterfly goes winging by;
A singing bird comes after;
And Nature, all from earth to sky,
Is bubbling o'er with laughter...

The earth is just so full of fun
It really can't contain it;
And streams of mirth so freely run
The heavens seem to rain it...

Why, it's the climax of the year,—
The highest time of living!—
Till naturally its bursting cheer
Just melts into thanksgiving.

8. Read the excerpt from the poem "Merry Autumn" by Paul Laurence Dunbar, then answer the question.

The phrases "purple tints", "modest green", "singing bird", "bubbling o'er with laughter" are examples of...

Ⓐ personification
Ⓑ sensory details
Ⓒ metaphors
Ⓓ onomatopoeia

9. Read the excerpt from the poem "Merry Autumn" by Paul Laurence Dunbar, then answer the question. Which line from the poem best captures the author's main idea?

Ⓐ "Now purple tints are all around"
Ⓑ "And Nature, all from earth to sky/Is bubbling o'er with laughter"
Ⓒ "Why, it's the climax of the year/The highest time of living"
Ⓓ "Till naturally its busting cheer/Just melts into thanksgiving"

10. Read the sentence. Based on the context, determine the meaning of the underlined word.

"Zachary expected to be punished severely, which was why he was surprised when his parents offered him <u>clemency</u>."

Ⓐ mercy
Ⓑ an appeal
Ⓒ defense
Ⓓ choice

DAY 5

CHALLENGE YOURSELF!

✓ Equation of a Circle
✓ Reading-Writing-Language

www.lumoslearning.com/a/dc10-30

See the first page for Signup details

WEEK 6 - ESSAY PROMPT

✓ What Captivates You?

www.lumoslearning.com/a/wc10-w6

See the first page for Signup details

Stop Making These Silly Mistakes on the SAT/ACT

Writing Mistakes

Assuming You Know The Grammar

It can be very tempting to "listen" to what sentence sounds better, but many grammatical questions rely on a specific and standardized set of grammar rules. It is crucial that you are familiar with the grammar before the exam, as it may not be as intuitive as you would hope.

Choosing The Answer Too Quickly

In multiple-choice questions, there will often be an option that is almost correct. This option can easily trick you into haphazardly selecting the wrong answer. Make sure not to rush and read every option thoroughly; sometimes, the difference between correct and incorrect is subtle.

Reading Mistakes

Losing Track Of Time

The reading portion of the tests can be very fast-paced, requiring you to read and answer each passage extremely quickly. By dwelling on a particular passage or question, you may run out of time. Instead, if you are unsure, mark the question and come back to it later.

Relying On Previous Knowledge

By this point, you probably have a thorough understanding of many broad concepts and contexts of the world in general. This is an excellent thing but can get in the way of answering exam questions succinctly. Most questions will provide the contextual clues needed to answer correctly and should not require much prior knowledge. If you are answering based on the information you never studied, you're likely going in the wrong direction.

Maths Mistakes

Forgetting Formulas

While some formulas will be provided, there is a chance that you will need certain formulas that aren't presented with the exam. Besides, checking formulas takes time and disrupts that important exam momentum. Include a formula sheet in your study, and spend some time drilling yourself on them.

Skipping Details

In a rush to finish your exam, it can be easy to brush oversteps in a mathematical equation. Especially in non-calculator sections, you could easily make a mistake that results in the wrong answer by trying to do it all in your head. Writing down every step also makes it easier to check your work and troubleshoot if you have any spare time in the end.

There's no need to panic- if you take the time to prepare for your upcoming SAT/ACT you can be calm and ready. Many of these little mistakes can easily be eliminated through practice, so scheduling in some practice exams is a crucial step to acing the test. On test day, try to keep a clear head, and you will surely avoid any unforced errors.

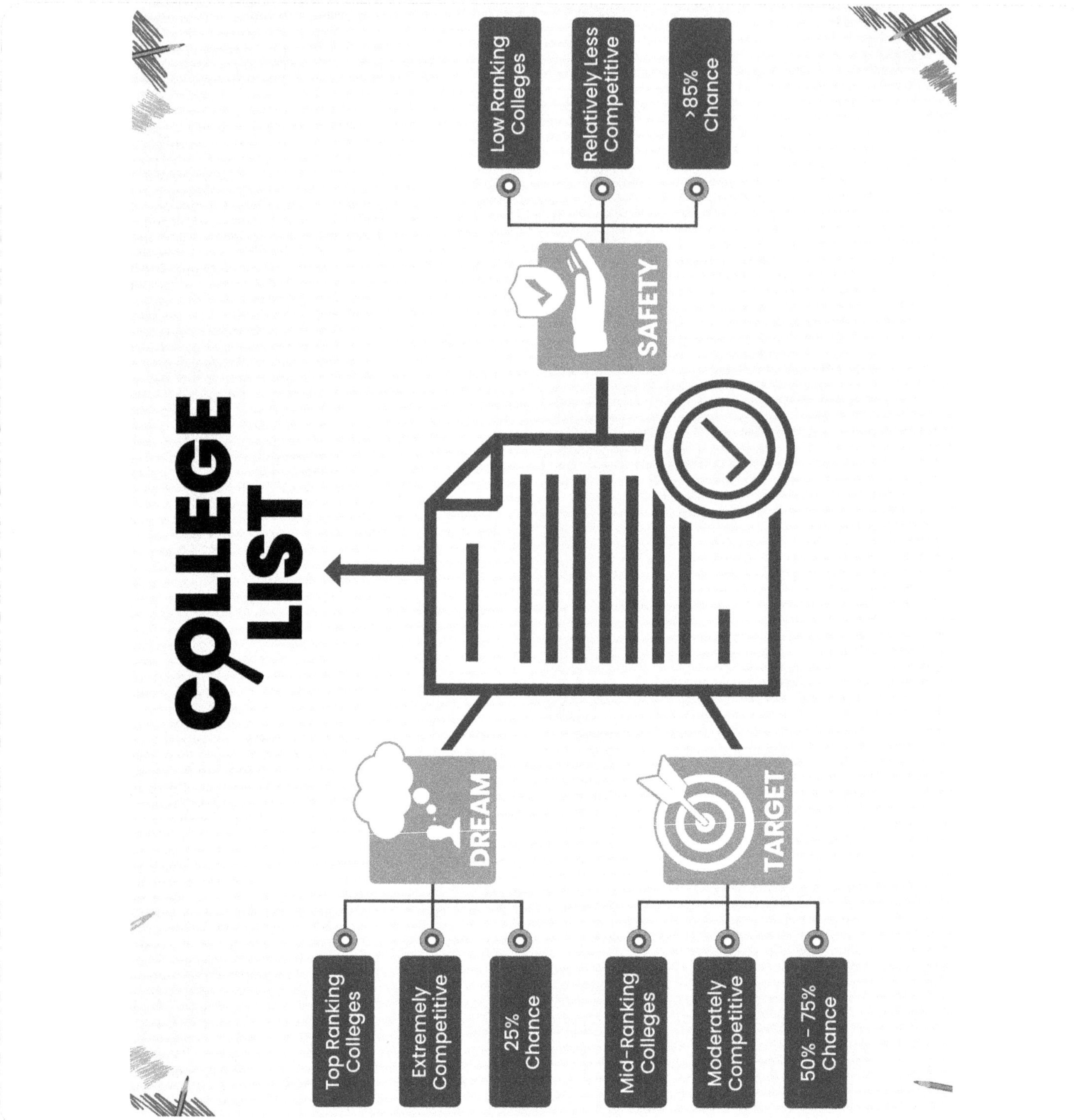

COLLEGE READINESS MIND MAPS

✔ College List

www.lumoslearning.com/a/crm6

See the first page for Signup details

WEEK 7
SUMMER PRACTICE

SOLVE SYSTEMS OF LINEAR EQUATIONS EXACTLY AND APPROXIMATELY

DAY 1

1. Solve the system of equations by graphing the lines.

x-y=3
7x-y=-3

Ⓐ (-1,-3)
Ⓑ (-1,-4)
Ⓒ (-2,-4)
Ⓓ (-2,-5)

2. Solve the system of equations by graphing the lines.

y=-2x+1
y=x-5

Ⓐ (2,-3)
Ⓑ (1,-4)
Ⓒ (3,-4)
Ⓓ (2,-5)

3. Solve the system of equations by graphing the lines.

3x-y=4
x+2y=6

Ⓐ (-1,-3)
Ⓑ (3,2)
Ⓒ (2,3)
Ⓓ (2,2)

4. Solve the system of equations by graphing the lines.

x−y=-2
x=−3

Ⓐ (−1,−3)
Ⓑ (−3,−1)
Ⓒ (−4,−1)
Ⓓ (−2,−1)

5. Two numbers have a sum of 24 and a difference of 6. Write a linear system of equations and solve to find the two numbers.

Ⓐ (10, 14)
Ⓑ (15, 9)
Ⓒ No solution
Ⓓ (9, 15)

The Tell-Tale Heart

-by Edgar Allan Poe

Read the passage below and answer the questions that follow.

TRUE! -- nervous -- very, very dreadfully nervous I had been and am; but why will you say that I am mad? The disease had sharpened my senses -- not destroyed -- not dulled them. Above all was the sense of hearing acute. I heard all things in the heaven and in the earth. I heard many things in hell. How, then, am I mad? Hearken! and observe how healthily -- how calmly I can tell you the whole story.

It is impossible to say how first the idea entered my brain; but once conceived, it haunted me day and night. Object there was none. Passion there was none. I loved the old man. He had never wronged me. He had never given me insult. For his gold I had no desire. I think it was his eye! yes, it was this! He had the eye of a vulture --a pale blue eye, with a film over it. Whenever it fell upon me, my blood ran cold; and so by degrees -- very gradually --I made up my mind to take the life of the old man, and thus rid myself of the eye forever.

Now this is the point. You fancy me mad. Madmen know nothing. But you should have seen me. You should have seen how wisely I proceeded --with what caution --with what foresight --with what dissimulation I went to work! I was never kinder to the old man than during the whole week before I killed him. And every night, about midnight, I turned the latch of his door and opened it --oh so gently! And then, when I had made an opening sufficient for my head, I put in a dark lantern, all closed, closed, so that no light shone out, and then I thrust in my head. Oh, you would have laughed to see how cunningly I thrust it in! I moved it slowly --very, very slowly, so that I might not disturb the old man's sleep. It took me an hour to place my whole head within the opening so far that I could see him as he lay upon his bed. Ha! --would a madman have been so wise as this? And then, when my head was well in the room, I undid the lantern cautiously --oh, so cautiously --cautiously (for the hinges creaked) --I undid it just so much that a single thin ray fell upon the vulture eye. And this I did for seven long nights --every night just at midnight --but I found the eye always closed; and so it was impossible to do the work; for it was not the old man who vexed me, but his Evil Eye. And every morning, when the day broke, I went boldly into the chamber, and spoke courageously to him, calling him by name in a hearty tone, and inquiring how he had passed the night. So you see he would have been a very profound old man, indeed, to suspect that every night, just at twelve, I looked in upon him while he slept.

Upon the eighth night I was more than usually cautious in opening the door. A watch's minute hand moves more quickly than did mine. Never before that night had I felt the extent of my own powers --of my sagacity. I could scarcely contain my feelings of triumph. To think that there I was, opening the door, little by little, and he not even to dream of my secret deeds or thoughts. I fairly chuckled at the idea; and perhaps he heard me; for he moved on the bed suddenly, as if startled. Now you may think that I drew back --but no. His room was as black as pitch with the thick darkness, (for the shutters were close fastened, through fear of robbers,) and so I knew that he could not see the opening of the door, and I kept pushing it on steadily,

steadily.

I had my head in, and was about to open the lantern, when my thumb slipped upon the tin fastening, and the old man sprang up in bed, crying out --"Who's there?"

I kept quite still and said nothing. For a whole hour I did not move a muscle, and in the meantime I did not hear him lie down. He was still sitting up in the bed listening; --just as I have done, night after night, hearkening to the death watches in the wall.

Presently I heard a slight groan, and I knew it was the groan of mortal terror. It was not a groan of pain or of grief --oh, no! --it was the low stifled sound that arises from the bottom of the soul when overcharged with awe. I knew the sound well. Many a night, just at midnight, when all the world slept, it has welled up from my own bosom, deepening, with its dreadful echo, the terrors that distracted me. I say I knew it well. I knew what the old man felt, and pitied him, although I chuckled at heart. I knew that he had been lying awake ever since the first slight noise, when he had turned in the bed. His fears had been ever since growing upon him. He had been trying to fancy them causeless, but could not. He had been saying to himself --"It is nothing but the wind in the chimney --it is only a mouse crossing the floor," or "It is merely a cricket which has made a single chirp." Yes, he had been trying to comfort himself with these suppositions: but he had found all in vain. All in vain; because Death, in approaching him had stalked with his black shadow before him, and enveloped the victim. And it was the mournful influence of the unperceived shadow that caused him to feel --although he neither saw nor heard --to feel the presence of my head within the room.

When I had waited a long time, very patiently, without hearing him lie down, I resolved to open a little --a very, very little crevice in the lantern. So I opened it --you cannot imagine how stealthily, stealthily -- until, at length a single dim ray, like the thread of the spider, shot from out the crevice and fell full upon the vulture eye.

It was open --wide, wide open --and I grew furious as I gazed upon it. I saw it with perfect distinctness --all a dull blue, with a h ideous veil over it that chilled the very marrow in my bones; but I could see nothing else of the old man's face or person: for I had directed the ray as if by instinct, precisely upon the damned spot.

And have I not told you that what you mistake for madness is but over acuteness of the senses? --now, I say, there came to my ears a low, dull, quick sound, such as a watch makes when enveloped in cotton.

I knew that sound well, too. It was the beating of the old man's heart. It increased my fury, as the beating of a drum stimulates the soldier into courage.

But even yet I refrained and kept still. I scarcely breathed. I held the lantern motionless. I tried how steadily I could maintain the ray upon the eye. Meantime the hellish tattoo of the heart increased. It grew quicker and quicker, and louder and louder every instant. The old man's terror must have been extreme! It grew louder, I say, louder every moment! --do you mark me well? I have told you that I am nervous: so I am. And now at the dead hour of the night, amid the dreadful silence of that old house, so strange a noise as this excited me to uncontrollable terror. Yet, for some minutes longer I refrained and stood still. But the beating grew louder, louder! I thought the heart must burst. And now a new anxiety seized me --the sound would be heard by a neighbor! The old man's hour had come! With a loud yell, I threw open the lantern and leaped into the room. He shrieked once --once only. In an instant I dragged him to the floor, and pulled the heavy bed over him. I then smiled gaily, to find the deed so far done.

But, for many minutes, the heart beat on with a muffled sound. This, however, did not vex me; it would not be heard through the wall. At length it ceased. The old man was dead. I removed the bed and examined the corpse. Yes, he was stone, stone dead. I placed my hand upon the heart and held it there many minutes. There was no pulsation. He was stone dead. His eye would trouble me no more.

If still you think me mad, you will think so no longer when I describe the wise precautions I took for the concealment of the body. The night waned, and I worked hastily, but in silence. First of all I dismembered the corpse. I cut off the head and the arms and the legs.

I then took up three planks from the flooring of the chamber, and deposited all between the scantlings. I then replaced the boards so cleverly, so cunningly, that no human eye --not even his --could have detected any thing wrong. There was nothing to wash out --no stain of any kind --no blood-spot whatever. I had been too wary for that. A tub had caught all --ha! ha!

When I had made an end of these labors, it was four o'clock --still dark as midnight. As the bell sounded the hour, there came a knocking at the street door. I went down to open it with a light heart, --for what had I now to fear? There entered three men, who introduced themselves, with perfect suavity, as officers of the police. A shriek had been heard by a neighbor during the night; suspicion of foul play had been aroused; information had been lodged at the police office, and they (the officers) had been deputed to search the premises.

I smiled, --for what had I to fear? I bade the gentlemen welcome. The shriek, I said, was my own in a dream. The old man, I mentioned, was absent in the country. I took my visitors all over the house. I bade them search --search well. I led them, at length, to his chamber. I showed them his treasures, secure, undisturbed. In the enthusiasm of my confidence, I brought chairs into the room, and desired them here to rest from their fatigues, while I myself, in the wild audacity of my perfect triumph, placed my own seat upon the very spot beneath which reposed the corpse of the victim.

The officers were satisfied. My manner had convinced them. I was singularly at ease. They sat, and while I answered cheerily, they chatted of familiar things. But, ere long, I felt myself getting pale and wished them gone. My head ached, and I fancied a ringing in my ears: but still they sat and still chatted. The ringing became more distinct: --it continued and became more distinct: I talked more freely to get rid of the feeling: but it continued and gained definiteness --until, at length, I found that the noise was not within my ears.

No doubt I now grew very pale; --but I talked more fluently, and with a heightened voice. Yet the sound increased --and what could I do? It was a low, dull, quick sound --much such a sound as a watch makes when enveloped in cotton. I gasped for breath --and yet the officers heard it not. I talked more quickly --more vehemently; but the noise steadily increased. I arose and argued about trifles, in a high key and with violent gesticulations; but the noise steadily increased. Why would they not be gone? I paced the floor to and fro with heavy strides, as if excited to fury by the observations of the men -- but the noise steadily increased. Oh God! what could I do? I foamed --I raved --I swore! I swung the chair upon which I had been sitting, and grated it upon the boards, but the noise arose over all and continually increased. It grew louder --louder --louder! And still the men chatted pleasantly, and smiled. Was it possible they heard not? Almighty God! --no, no! They heard! --they suspected! --they knew! --they were making a mockery of my horror! --this I thought, and this I think. But anything was better than this agony! Anything was more tolerable than this derision! I could bear those hypocritical smiles no longer! I felt that I must scream or die! --and now --again! --hark! louder! louder! louder! louder! --

"Villains!" I shrieked, "dissemble no more! I admit the deed! --tear up the planks! --here, here! --it is the beating of his hideous heart!"

6. From the beginning of the passage, it is reasonable to infer that:

Ⓐ the narrator is worried that he might be crazy
Ⓑ the narrator is confident of his sanity
Ⓒ the narrator is angered at the accusation of madness
Ⓓ the narrator is convinced of his own madness

7. Choose the statement that best reflects the theme of the story.

Ⓐ People will always uncover the secrets we try to hide.
Ⓑ The cost of murder is the perpetrator's sanity.
Ⓒ Madness is the result of suffering another person's imperfections.
Ⓓ When guilt becomes overwhelming, the truth comes out.

8. The narrator equates sanity to ____.

Ⓐ His love for the old man.
Ⓑ His attention to detail in committing the crime.
Ⓒ His ability to hear a sound that other people couldn't.
Ⓓ His nervousness at being accused of madness.

9. What is the impact of the author using an unreliable narrator?

Ⓐ The credibility of the story is in question.
Ⓑ The narrator includes greater detail in his telling of the story.
Ⓒ The accuracy of the story is maintained.
Ⓓ The plot of the story is confusing.

10. Which of the following synonyms is closest to the meaning of the word "fancy" as it is used in paragraph 3?

Ⓐ elaborate
Ⓑ believe
Ⓒ craving
Ⓓ wish

1. Given that $i^2 = -1$, what is i^{34}?

Ⓐ i
Ⓑ -1
Ⓒ -i
Ⓓ 1

2. Given that $i^2 = -1$, what is $\sqrt{49} - \sqrt{-9}$?

Ⓐ -21
Ⓑ -21i
Ⓒ 21i
Ⓓ 21

3. Given that $i^2 = -1$, what is $7.(-6)i^{137}$?

Ⓐ 42i
Ⓑ -42
Ⓒ -42i
Ⓓ 42

4. Why does $\sqrt{-1}$ need to be imaginary?

Ⓐ Because $(1)^2=1$ and $(-1)^2=1$
Ⓑ Because -1 is its own reciprocal
Ⓒ Because real numbers are not imaginary
Ⓓ Because imaginary people need numbers too

5. What does i^2 simplify to be?

Ⓐ -i
Ⓑ i
Ⓒ 1
Ⓓ -1

6. Each resource below would be helpful when researching the effects of pollution on our environment, except:

Ⓐ an article from a science journal
Ⓑ an Internet video made by a biology student
Ⓒ a website on global warming
Ⓓ none of these

7. Which website might be helpful for a research project on the death penalty?

Ⓐ prodeathpenalty.com
Ⓑ https://www.scu.edu/ethics
Ⓒ bbc.com
Ⓓ amnestyinternational.org

8. Which question would not be helpful when researching a project on female leaders?

Ⓐ How have our male leaders shaped the role for women?
Ⓑ How have female leaders had the most impact on society?
Ⓒ What roles have these women played in society?
Ⓓ What influences have had the biggest impact on specific female leaders?

9. Determine whether the following statement is True or False.

There is no need for print sources, and it is appropriate to use only the Internet in your research.

Ⓐ True
Ⓑ False

10. Based on the prompt "Prepare a report about something you are interested in," what is the process for selecting a good research topic?

Ⓐ Identify 2 or 3 interests and prepare your index cards for researching those topics.
Ⓑ Identify 2 or 3 interests, and determine which is researchable and most interesting to you, and then formulate a question about that interest.
Ⓒ Tell the teacher that you have no interests and don't know what to write about.
Ⓓ Start writing about what you like and research after.

DAY 2

CHALLENGE YOURSELF!

✔ Introduction to Imaginary and Complex Numbers
✔ Writing

www.lumoslearning.com/a/dc10-32

See the first page for Signup details

ADD, SUBTRACT, AND MULTIPLY COMPLEX NUMBERS

DAY 3

1. Given that $i^2 = -1$, what is $7i^{127}.(-6)i^{138}$?

Ⓐ $42i$
Ⓑ -42
Ⓒ $-42i$
Ⓓ 42

2. Given that $i^2 = -1$, what is $(9+4i) + (-6-2i)$?

Ⓐ $3-2i$
Ⓑ $3+2i$
Ⓒ $15+6i$
Ⓓ $13-8i$

3. Given that $i^2 = -1$, what is $(6-7i)\times(4-5i)$?

Ⓐ $11-58i$
Ⓑ $-11-58i$
Ⓒ $11+58i$
Ⓓ $-11+58i$

4. Find the product of 5 and $2-3i$?

Ⓐ $10+2i$
Ⓑ $7+2i$
Ⓒ $10-15i$
Ⓓ $10+15i$

5. Find the product of $2+4i$ and $1-i$?

Ⓐ $2+2i+4i^2$
Ⓑ $2-4i^2$
Ⓒ 6
Ⓓ $6+2i$

6. Which of the following words contains a positive connotation for protest?

Ⓐ disturbance
Ⓑ uprising
Ⓒ riot
Ⓓ demonstration

7. Which figure of speech is being used in the highlighted part of this text?

"I'm so hungry I could eat a horse," Stan complained.

Ⓐ euphemism
Ⓑ hyperbole
Ⓒ alliteration
Ⓓ irony

8. Which figure of speech is being used in the highlighted part of this text?

"After her dog died, a dark cloud hung over her head, and she moped around in sorrow."

Ⓐ personification
Ⓑ hyperbole
Ⓒ simile
Ⓓ metaphor

9. Which of the following carries a negative connotation?

Ⓐ student
Ⓑ learner
Ⓒ nerd
Ⓓ apprentice

10. When an author wants to replace an unpleasant word or expression with something more mild or indirect, it is called a(n):

Ⓐ alliteration
Ⓑ figure of speech
Ⓒ idiom
Ⓓ euphemism

1. Which function is graphed below?

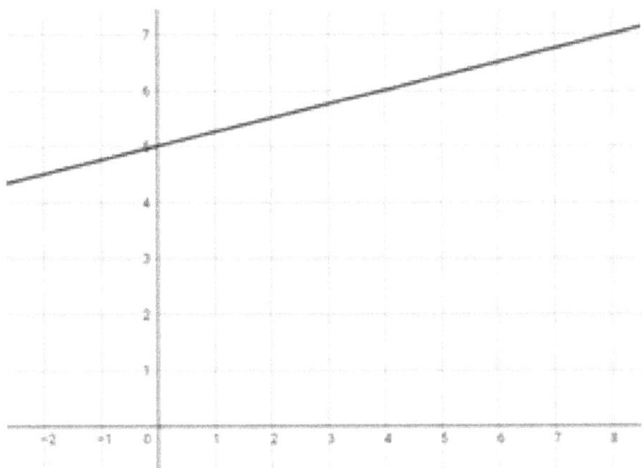

Ⓐ $f(y)=\frac{1}{4}x+4$

Ⓑ $f(y)=\frac{1}{4}y+4$

Ⓒ $f(x)=4x+4$

Ⓓ $f(x)=4x+\frac{1}{4}$

2. Which arithmetic sequence rule produces the term values shown in the table below?

n	1	2	3	4
a	5	10	15	20

Ⓐ $a_n=5n$

Ⓑ $a_n=10n-5$

Ⓒ $a_n=n+4$

Ⓓ $a_n=4n+3$

3. Which function is graphed below?

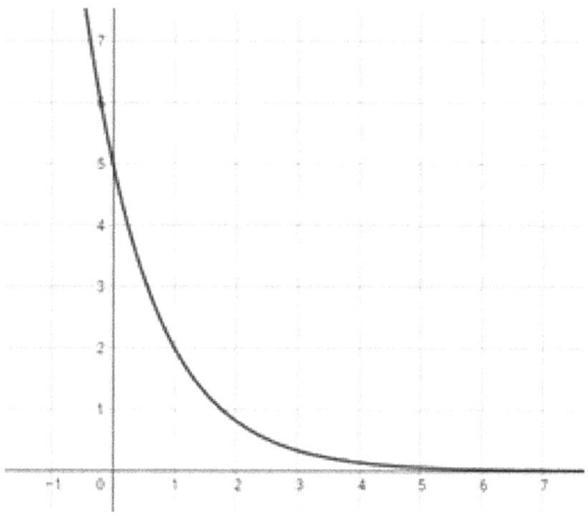

Ⓐ $f(x)=5(0.5)^x$
Ⓑ $f(x)=5(0.4)^x$
Ⓒ $f(x)=4(0.5)^x$
Ⓓ $f(x)=5(1.5)^x$

4. Which is the equation of the given graph?

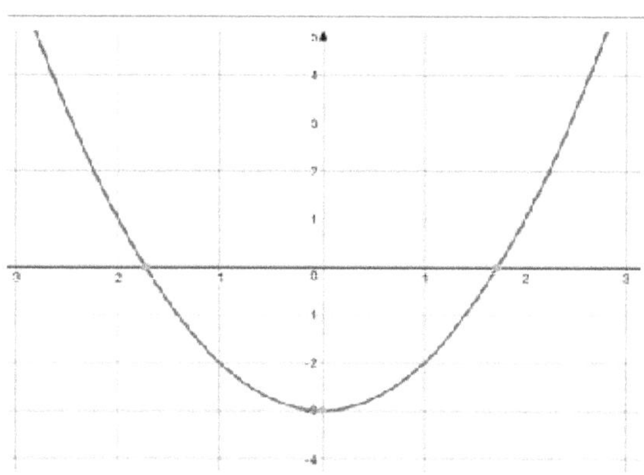

Ⓐ $y=x^2-3$
Ⓑ $y=x^2+3$
Ⓒ $y=3x^2$
Ⓓ $y=-3x^2$

5. Shelly decided to start a home based cookie baking business. During the first year she baked 20 dozen cookies, and the next year she baked 30 dozen. The year after that she produced 40 dozen. What function could model this success (assuming that x = 0 represents the first year)?

Ⓐ $f(x)=20x$
Ⓑ $f(x)=10x+20$
Ⓒ $f(x)=10x$
Ⓓ None of these

President Lincoln's Speech

Read the passage below and answer the questions that follow.

"Fourscore and seven years ago our fathers brought forth, on this continent, a new nation, conceived in liberty, and dedicated to the proposition that all men are created equal. Now we are engaged in a great civil war, testing whether that nation, or any nation so conceived, and so dedicated, can long endure. We are met on a great battle-field of that war. We have come to dedicate a portion of that field, as a final resting-place for those who here gave their lives, that that nation might live. It is altogether fitting and proper that we should do this. But, in a larger sense, we cannot dedicate, we cannot consecrate—we cannot hallow—this ground. The brave men, living and dead, who struggled here, have consecrated it far above our poor power to add or detract.

6. The purpose of this speech is to:

Ⓐ Respectfully dedicate a portion of the ground for the soldiers who have died in battle.
Ⓑ Emphasize Lincoln's anger at the American people.
Ⓒ Question whether or not America is going to make it or if it will self-destruct.
Ⓓ Both B and C

7. The tone of Lincoln's speech is:

Ⓐ Happy
Ⓑ Worried
Ⓒ Respectful
Ⓓ Both B and C

8. Which rhetorical device is not used in Lincoln's Inaugural address?

Ⓐ Allusion
Ⓑ Contrast
Ⓒ Metaphor
Ⓓ Call to action

9. Lincoln's main purpose in this excerpt is to ______ his audience

Ⓐ Unite
Ⓑ Scare
Ⓒ Anger
Ⓓ Subdue

Constitutional Plans

Read the text below and answer the question that follows.

At the Constitutional Convention, there was a debate over two plans for setting up the legislative branch of government. These were the two plans:

VIRGINIA PLAN
Each state would be represented according to the size of its population.

NEW JERSEY PLAN
Each state would have the same number of representatives.
The debate was resolved by a third plan:

CONNECTICUT PLAN
(the GREAT COMPROMISE)
The legislature would have two houses. In one house the states would be represented according to population. In the other house each state would be represented equally.

10. What issue was settled by the Connecticut Plan?

Ⓐ how to count the states' population
Ⓑ how the president should be elected
Ⓒ how the larger and smaller states could share power
Ⓓ how the states and the federal government could share power

EQUATION OF A CIRCLE

DAY 5

1. If a circle has an equation of $(x+1)^2+(y-7)^2=23$, what are the coordinates of the center and the value of the radius?

Ⓐ $C(1,-7);r=\sqrt{23}$
Ⓑ $C(1,49);r=23$
Ⓒ $C(-1,7);r=\sqrt{23}$
Ⓓ $C(-1,7);r=23$

2. A circle has its center at (-8, 1) and has a diameter of 10. Find the equation of the circle.

Ⓐ $(x+8)^2+(y-1)^2=100$
Ⓑ $(x+8)^2+(y-1)^2=25$
Ⓒ $(x-8)^2+(y+1)^2=25$
Ⓓ $(x-8)^2+(y+1)^2=100$

3. The graph of $x^2+y^2=49$ is translated 12 units left and 9 units down. What is the equation of the new circle?

Ⓐ $(x-12)^2+(y-9)^2=49$
Ⓑ $(x+6)^2+(y+3)^2=49$
Ⓒ $(x+12)^2+(y+9)^2=49$
Ⓓ $(x-6)^2+(y-3)^2=49$

4. What is the equation of a circle, in the form, $(x-h)^2+(y-k)^2=r^2$, if the center of the circle is at the point (2,-1) and the point (4,2) is on the circle?

Ⓐ $(x+2)^2+(y-1)^2=13$
Ⓑ $(x+2)^2+(y-1)^2=\sqrt{13}$
Ⓒ $(x-2)^2+(y+1)^2=13$
Ⓓ $(x-2)^2+(y+1)^2=\sqrt{13}$

5. What is the equation of a circle, in the form, $(x-h)^2+(y-k)^2=r^2$, if the center of the circle is at the point (0,3) and the point (3,7) is on the circle?

Ⓐ $x^2+(y-3)^2=25$
Ⓑ $x^2+(y-3)^2=5$
Ⓒ $(x-3)^2+y^2=25$
Ⓓ $x^2+(y+3)^2=5$

6. Why is it important to create a research question as one of the first steps in the writing process?

Ⓐ It helps narrow down what is being researched and lays the groundwork for what will be researched.
Ⓑ It helps the researcher to group and categorize different aspects of their thesis.
Ⓒ It sets up the structure of the paper.
Ⓓ It helps you gather the quotes for your essay.

7. Which of the following sources does NOT need to be included in your bibliography or works cited page?

Ⓐ Title of a magazine
Ⓑ Author's Name
Ⓒ Publishing House
Ⓓ None of the above

8. Determine whether the following statement is True or False.

When searching for information on your topic, you should focus exclusively on .org, .edu, and .gov sites, because other domains can never be credible.

Ⓐ True
Ⓑ False

Merry Autumn

- by Paul Laurence Dunbar

Read the poem below and answer the questions that follow.

Now purple tints are all around;
The sky is blue and mellow;
And e'en the grasses turn the ground
From modest green to yellow...

A butterfly goes winging by;
A singing bird comes after;
And Nature, all from earth to sky,
Is bubbling o'er with laughter...

The earth is just so full of fun
It really can't contain it;
And streams of mirth so freely run
The heavens seem to rain it...

Why, it's the climax of the year,—
The highest time of living!—
Till naturally its bursting cheer
Just melts into thanksgiving.

9. Read the excerpt from the poem "Merry Autumn" by Paul Laurence Dunbar, then answer the question.

The phrases "purple tints", "modest green", "singing bird", "bubbling o'er with laughter" are examples of...

Ⓐ personification
Ⓑ sensory details
Ⓒ metaphors
Ⓓ onomatopoeia

10. Read the excerpt from the poem "Merry Autumn" by Paul Laurence Dunbar, then answer the question.

Which line from the poem best captures the author's main idea?

Ⓐ "Now purple tints are all around"
Ⓑ "And Nature, all from earth to sky/Is bubbling o'er with laughter"
Ⓒ "Why, it's the climax of the year/The highest time of living"
Ⓓ "Till naturally its busting cheer/Just melts into thanksgiving"

DAY 5

CHALLENGE YOURSELF!

✓ Equation of a Circle
✓ Writing

www.lumoslearning.com/a/dc10-35

See the first page for Signup details

WEEK 7 - ESSAY PROMPT

✓ Gratitude to Motivation

www.lumoslearning.com/a/wc10-w7

See the first page for Signup details

10 Actionable Tips to Ace Your Upcoming SAT

Keep Your Cool

While nerves are normal, managing SAT day anxiety is crucial in keeping a level head. A great way to do this is to be as familiar as possible with the structure and timing of the test... that's right, practice tests. The exam won't feel as daunting if you've done practice tests under pressure before.

Have A Plan For Each Section

When you take your practice tests, you will realize that each section has its quirks. Becoming familiar with the structure of the Reading, Writing, and Language, and Maths tests will help you plan your exam strategy.

Know Your Strengths

Everyone has strengths and weaknesses; that's just a fact. By identifying where you are strongest you can anticipate where you can get easy points, as well as plan where you'll spend more time.

Don't Do The Questions In Order

Do the easy questions first-you will build your confidence and leave more time for problem-solving the difficult questions.

Read Instructions Clearly

It seems simple, but a lot of context can be found in the questions! Make sure not to skip over important details.

Eliminate The Wrong Answers

With multiple-choice questions, the process of elimination is crucial. As long as you can confidently cross out all but one wrong answer, you will have the correct answer.

Memorize Grammar Rules

The SAT uses a standardized and specific set of grammatical rules. Become familiar with these, as the correct grammar is the basis for excellent writing.

Memorize Mathematical Formulas

While you will have access to formulas on test day, memorizing them is a great way to keep momentum and save time. You will probably commit many of them to memory during your study, but it can't hurt to actively memorize the most common mathematical formulas.

Double Check Your Answers

Another simple trick – Double-checking is an important step to avoid silly mistakes. If you go back at the end of each section, you may catch errors you previously missed or suddenly have an answer come to you. This is crucial for quality control.

Prepare The Night Before

The night before the SAT, try to avoid cramming in the study. Instead, prepare your stationery and calculator, eat a proper meal, and get a full night's sleep. Anything you haven't studied by now is unlikely to stick, and it is more important to keep a clear and alert mind.

The Bottom Line

Preparing for college can be a daunting task for students who are still just on the verge of adulthood. It is important to help prepare them as early as possible to be ready and confident for what is to come.

LETTER OF RECOMMENDATION

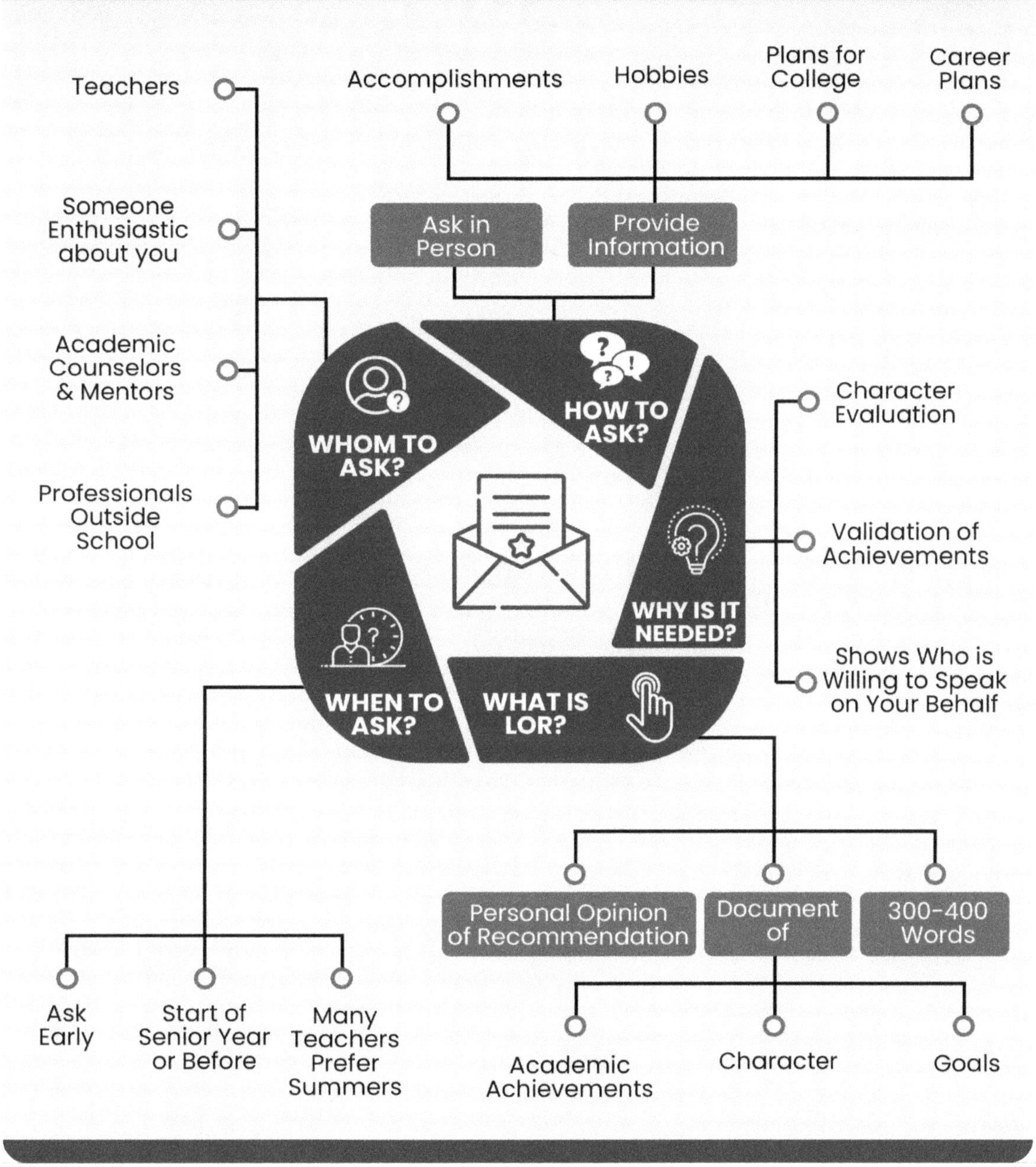

COLLEGE READINESS MIND MAPS

✔ Letter of Recommendation

www.lumoslearning.com/a/crm7

See the first page for Signup details

WEEK 8
SUMMER PRACTICE

SOLVE A SIMPLE SYSTEM CONSISTING OF A LINEAR EQUATION AND A QUADRATIC EQUATION

DAY 1

1. Solve this system of equations algebraically.

$$y=-2$$
$$x^2+y^2=16$$

Ⓐ $(2\sqrt{3},-2)$
Ⓑ $(-2\sqrt{3},-2)$
Ⓒ $(\pm 2\sqrt{3},-2)$
Ⓓ $(\pm 12,-2)$

2. Solve this system of equations by graphing the system.
NOTE: The values in the choices are rounded to the nearest two decimal places.

$$y=x^2-8x+2$$
$$x+y=4$$

Ⓐ $(-1.27,4.27),(7.27,-3.27)$
Ⓑ $(-0.27,4.27),(7.27,-3.27)$
Ⓒ $(-0.27,4.27),(6.27,-3.27)$
Ⓓ $(-0.27,4.27),(7.27,-2.27)$

3. Solve this system of equations algebraically.

$$x=2$$
$$x^2+y^2=25$$

Ⓐ $(\pm\sqrt{21},2)$
Ⓑ $(2,\pm\sqrt{21})$
Ⓒ $(2,\sqrt{21})$
Ⓓ $(2,-\sqrt{21})$

4. Solve this system of equations algebraically.

$x^2+y^2=36$
$x=4$

Ⓐ $(4,2\sqrt{5})$
Ⓑ $(4,\pm 2\sqrt{5})$
Ⓒ $(\pm 2\sqrt{5},4)$
Ⓓ $(4,-2\sqrt{5})$

5. Solve this system of equations algebraically.

$x^2+y^2=9$
$y=3$

Ⓐ $(0,-3)$
Ⓑ $(0,3)$
Ⓒ $(0,\pm 3)$
Ⓓ $(\pm 3,0)$

Richard Cory

BY EDWIN ARLINGTON ROBINSON

Read the poem below and answer the questions that follow.

Whenever Richard Cory went down town,
We people on the pavement looked at him:
He was a gentleman from sole to crown,
Clean favored, and imperially slim.

And he was always quietly arrayed,
And he was always human when he talked;
But still he fluttered pulses when he said,
"Good-morning," and he glittered when he walked.

And he was rich—yes, richer than a king—
And admirably schooled in every grace:
In fine, we thought that he was everything
To make us wish that we were in his place.

So on we worked, and waited for the light,
And went without the meat, and cursed the bread;
And Richard Cory, one calm summer night,
Went home and put a bullet through his head.

6. In the poem 'Richard Cory,' the primary conflict is

Ⓐ Man vs. Man
Ⓑ Man vs. Self
Ⓒ Man vs. Thing
Ⓓ Man vs. Machine

7. The climax of the poem is

Ⓐ when Richard walks through town.
Ⓑ when Richard says "Good morning".
Ⓒ when Richard is described as rich.
Ⓓ when Richard shoots himself.

8. Which line best signifies how superior Richard Cory is to the townspeople?

Ⓐ We people on the pavement looked at him:
Ⓑ "Good-morning," and he glittered when he walked.
Ⓒ And he was rich—yes, richer than a king—
Ⓓ All of the above.

9. How does the last stanza contribute to the theme of the poem?

Ⓐ It suggests that money is the answer to everything
Ⓑ It implies that wealth leads to destruction
Ⓒ It supports the idea that money does not buy happiness
Ⓓ It shows the value of friendship and camaraderie

10. What literary device does the speaker use when he says, "And went without the meat, and cursed the bread"?

Ⓐ Mood
Ⓑ Foreshadowing
Ⓒ Alliteration
Ⓓ Symbolism

UNDERSTANDING THE RELATIONSHIP BETWEEN EQUATIONS & GRAPHS

DAY 2

1. Which point is on the graph of $f(x)= \frac{4}{5}x - \frac{9}{5}$?

Ⓐ (7,4)
Ⓑ (9,5)
Ⓒ (6,3)
Ⓓ (13,9)

2. Which point is on the graph of $f(x)=-19x+45$?

Ⓐ (9,−126)
Ⓑ (7,−89)
Ⓒ (8,−113)
Ⓓ (12,−143)

3. Which point is on the graph of $f(x)= \frac{3}{7}x + \frac{15}{7}$?

Ⓐ $(5, \frac{22}{7})$
Ⓑ $(-3, -\frac{6}{7})$
Ⓒ $(-6, \frac{16}{7})$
Ⓓ $(-4, \frac{-3}{7})$

4. Samantha recently started baking and decorating cookies as a way to earn extra money. The equation y = 2x - 16 models her profit after selling x dozen cookies. What are the x and y coordinates of her break-even point?

Ⓐ (0, 8)
Ⓑ (8, 0)
Ⓒ (4, 0)
Ⓓ (0, -16)

5. A financial advisor has constructed a model to predict the total profits of a company within the next month. The relationship is given by the equation $y = x^2 - 10$ where y is the total profit and x is the number of days. When will the company start to turn a profit?

Ⓐ between day 3 and day 4
Ⓑ between day 2 and day 3
Ⓒ between day 9 and day 10
Ⓓ between day 10 and day 11

6. Which of the following must be included in a Works Cited?

Ⓐ Title of book/article/website
Ⓑ Author
Ⓒ Publication Date
Ⓓ All of the above

7. What might indicate that a source is not credible?

Ⓐ It contains a .com or .org in the link.
Ⓑ It cites its sources.
Ⓒ It contains data.
Ⓓ All of the above.

8. Which would be the next step once you have found a source for your research?

Ⓐ Begin writing
Ⓑ Create an outline
Ⓒ Verify the credibility of the source
Ⓓ Get more sources

9. Which of the following would be an authoritative information source?

Ⓐ Can trust all websites
Ⓑ No websites can be trusted
Ⓒ .gov
Ⓓ .com

10. Which of the following shows a correct parenthetical citation?

Ⓐ Wordsworth stated that Romantic poetry was marked by a "spontaneous overflow of powerful feelings" (263)
Ⓑ The authors claim that surface reading looks at what is "evident, perceptible, apprehensible in texts" (Best and Marcus 9).
Ⓒ One online film critic stated that Fitzcarraldo "has become notorious for its near-failure and many obstacles" (Taylor, "Fitzcarraldo").
Ⓓ All of the above

1. The graph below represents the approximate drag on an aircraft while in flight, so that the drag on the aircraft is equal to a coefficient times the velocity cubed. Which statement is true about the origin on the graph?

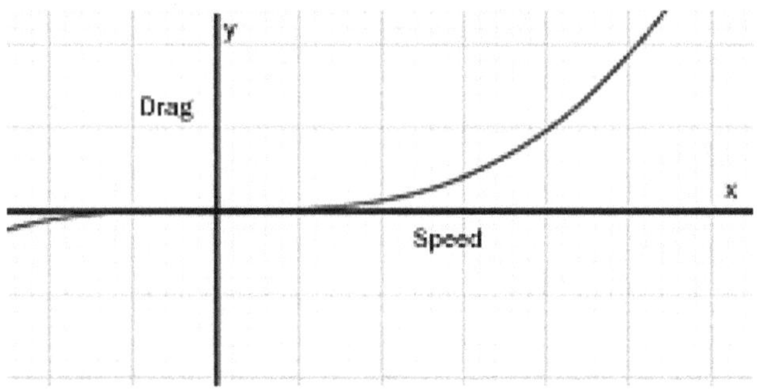

Ⓐ The origin shows the amount of drag after the airplane is in the air.
Ⓑ The origin shows that when the aircraft is not flying there is no drag.
Ⓒ The origin shows that the drag is increasing.
Ⓓ The origin shows that the drag changes from negative to positive.

2. Biologists discovered that the shoulder height of a male African elephant can be modeled by $h = 63.7\sqrt{t} + 83.5$, where h is the height in centimeters and t is the age in years. This function is graphed below. Which statement is true about the origin (0,0)?

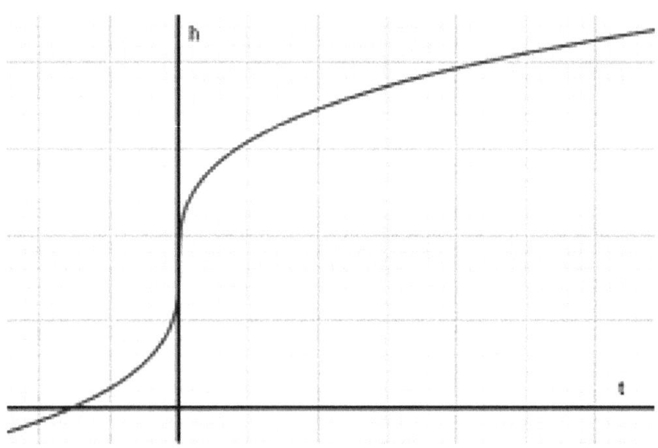

Ⓐ The origin shows the height of a male elephant at birth.
Ⓑ The graph does not pass through the origin because a male elephant is 83.5 cm tall at birth.
Ⓒ The origin represents the age of the elephant at birth.
Ⓓ The graph does not pass through the origin because a male elephant is 63.7 cm tall at birth.

3. The graph below shows the driving speed of a car over an 8 hour period during a mileage test. The speed limit on the freeway is universally 70 miles per hour, except when passing through cities when the speed limit is reduced. The driver of the car obeyed all speed limits during the test. Which statement is true about the size of each interval (each mark) on the y-axis of the graph?

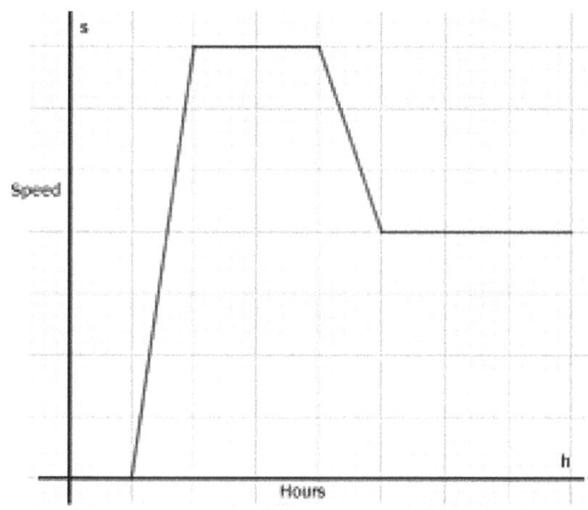

Ⓐ Each mark on the axis is approximately 10 miles per hour.
Ⓑ Each mark on the axis is approximately 7 miles per hour.
Ⓒ Each mark on the axis is approximately 12 miles per hour.
Ⓓ Each mark on the axis is approximately 9 miles per hour.

4. The graph below shows that the distance to a ship on the horizon is directly proportional to the square root of the elevation of the observer. In this context, which statement is most accurate about the origin?

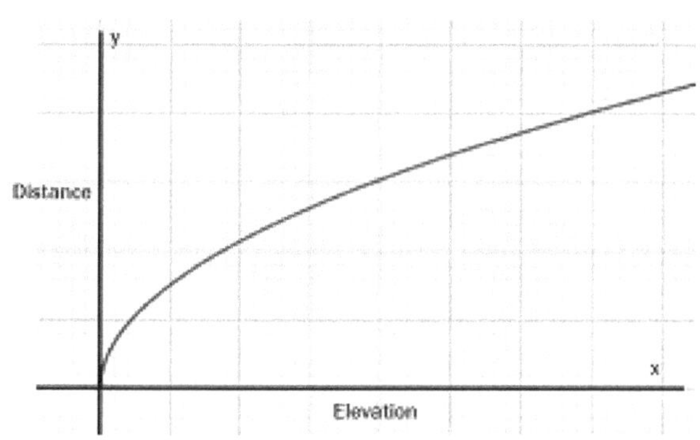

Ⓐ The origin shows that the person is not looking at the horizon.
Ⓑ The origin represents the elevation of the person.
Ⓒ The origin represents the distance from the person to the horizon.
Ⓓ The origin shows that if the person's elevation is at ground level, he/she cannot see any distance.

5. The graph represents the relationship between carat weight (x) and the number of diamonds purchased (y), modeled by a logarithmic function y = log(x + 1). Which statement best describes the graph?

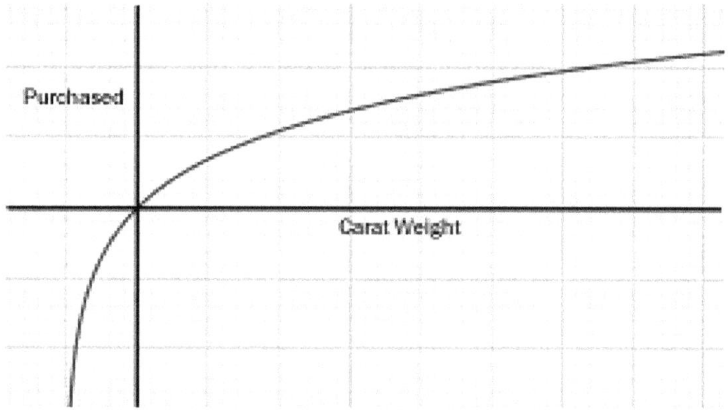

Ⓐ The graph shows 10 diamonds are purchased at 0 carats.
Ⓑ The number of diamonds purchased increases at a constant rate.
Ⓒ The number of diamonds purchased increases quickly at first, then more slowly.
Ⓓ The graph is symmetric about the y-axis.

6. How many sources should be listed in a Works Cited?

Ⓐ One
Ⓑ At least five
Ⓒ No limit
Ⓓ None of the above

7. An Earth Science teacher wants to take his students on a field trip to the local science museum. Which source is likely to be the most useful in planning this field trip?

Ⓐ A review from a local blogger describing her experience at the science museum with her two young children.
Ⓑ The city's tourist information website listing local attractions.
Ⓒ A list of all the places the school has taken students for field trips previously.
Ⓓ The science museum's official website.

8. Which is the correct way to order a Works Cited page in MLA format?

Ⓐ Alphabetized by author's last name.
Ⓑ In the order that the sources are cited in the paper.
Ⓒ In the order that the writer found and read each source.
Ⓓ By the date the source was published, from oldest to most recent.

9. A student is researching the opioid epidemic in America in an attempt to bring awareness to his classmates. He hopes to provide detailed information on how the drug problem affects Americans. Which of the following sources of information is likely to be the most credible?

Ⓐ An interview with the school nurse about the risk factors for students
Ⓑ Statistics of drug-related deaths from the US Department of Health and Human Services
Ⓒ Information about psychological effects of drugs found on an online forum
Ⓓ A drug counselling brochure provided by a local pharm

10. What is the correct way to cite a source within a paper when the source has no author listed?

Ⓐ Use "Anonymous" in place of the author's name.
Ⓑ Use the title of the source in place of the author's name.
Ⓒ Use your own name in place of the author's name.
Ⓓ Leave out the author's name and use the page number by itself.

1. Suppose there are 950 widgets in a warehouse when a new business opened. The business expects to sell 45 of the widgets per week. Which function I(w) represents the number of widgets that will be in the warehouse after selling them for w weeks?

Ⓐ I(w)=-950+45w
Ⓑ I(w)=950-45w
Ⓒ I(w)=950+45w
Ⓓ I(w)=45-950w

2. Suppose there are 1,090 hamburger patties in the walk-in cooler at a restaurant at the beginning of the week. The restaurant expects to sell 150 hamburger patties each day. Which function H(p) represents the number of hamburger patties that will be in the cooler after selling them for p days?

Ⓐ H(p)=1090-150p
Ⓑ H(p)=-1090+150p
Ⓒ H(p)=1090+150p
Ⓓ H(p)=1090p-150

3. Suppose a construction company is preparing to build a large apartment complex and is beginning to receive truckloads of sheets of plywood for the construction project. The company has 150 sheets on hand and each truckload has 80 sheets. Which function P(t) represents the total number of sheets of plywood the company has if t truckloads of plywood have been delivered?

Ⓐ P(t)=150-80t
Ⓑ P(t)=150+80t
Ⓒ P(t)=-150-80t
Ⓓ P(t)=-150 + 80t

4. The ABC Early Childhood Center will open for business with 60 students. Every year after opening the school expects to add 35 new students. Which function S(x) represents the number of students who will be enrolled at the school after x years?

Ⓐ S(x)=-60+35x
Ⓑ S(x)=60+35x
Ⓒ S(x)=60-35x
Ⓓ S(x)=60x+35

5. Suppose an amateur theater spent $495 to produce a play. If the tickets cost $15 each, which function P(t) represents the net profit of the play, with t representing the number of tickets sold?

Ⓐ $P(t)=15t + 495$
Ⓑ $P(t)=15t + 485$
Ⓒ $P(t)=15t - 495$
Ⓓ $P(t)=495t - 15$

6. If you wanted to find synonyms or antonyms for a word, what book would you consult?

Ⓐ Dictionary
Ⓑ Encyclopedia
Ⓒ Manual
Ⓓ Thesaurus

7. Choose the academic term that best completes the sentence. "China's ______ is largely dependent on manufacturing with 80% of their exports being manufactured goods."

Ⓐ government
Ⓑ research
Ⓒ bank
Ⓓ economy

8. Choose the academic term that best completes the sentence. "The main task of the ___ branch of the United States government is to make laws."

Ⓐ executive
Ⓑ legislative
Ⓒ congressional
Ⓓ judicial

9. Which of the following lines from "The Most Dangerous Game" contains a metaphor?

Ⓐ "The sea was as flat as a plate-glass window."
Ⓑ "Ten minutes of determined effort brought another sound to his ears--the most welcome he had ever heard--the muttering and growling of the sea breaking on a rocky shore."
Ⓒ "The lights of the yacht became faint and ever-vanishing fireflies; then they were blotted out entirely by the night."
Ⓓ "Bleak darkness was blacking out the sea and jungle when Rainsford sighted the lights."

10. Choose the academic term that best completes the sentence. "For teenagers, whose brains are not yet fully developed, the ___ of rational and irrational thinking are not always clear."

Ⓐ assets
Ⓑ concepts
Ⓒ roles
Ⓓ factors

1. A farmer wishes to enclose a square plot of land with a fence so that he can grow a garden. Find the amount of fencing required if the length of a side is 15 ft.

Ⓐ 45 ft
Ⓑ 30 ft
Ⓒ 225 ft
Ⓓ 60 ft

2. The sail on a boat is in the shape of a right triangle. If the legs are 29.5 feet and 10.5 feet, what is the area of the sail rounded to the nearest whole number?

Ⓐ 309 ft^2
Ⓑ 155 ft^2
Ⓒ 40 ft^2
Ⓓ 154 ft^2

3. The sail on a boat is in the shape of a right triangle. If the legs are 5 feet and 10 feet, what is the perimeter of the triangle rounded to the nearest tenth?

Ⓐ 140.3 ft
Ⓑ 20.6 ft
Ⓒ 26.2 ft
Ⓓ 11.8 ft

4. A circular tablecloth has a radius of 6 feet. Anna would like to put a border of ribbon around the entire tablecloth. How much ribbon does she need, rounded to the nearest hundredth of a foot?

Ⓐ 37.70 ft
Ⓑ 75.40 ft
Ⓒ 18.85 ft
Ⓓ 21.63 ft

5. The circular face of a watch has a diameter of 5 cm. What is the area of the face of the watch rounded to the nearest whole number?

Ⓐ 79 cm^2
Ⓑ 16 cm^2
Ⓒ 20 cm^2
Ⓓ 31 cm^2

6. When the audience knows something that a character in a play doesn't, it is called

Ⓐ situational irony
Ⓑ sarcasm
Ⓒ verbal irony
Ⓓ dramatic irony

7. Which figure of speech is being used in the highlighted part of this text?

"The pepper tickled her nose, causing her to sneeze right into her dish."

Ⓐ personification
Ⓑ hyperbole
Ⓒ simile
Ⓓ metaphor

8. Phrases such as "jumbo shrimp", "minor crisis", and "awfully good" are examples of what figure of speech?

Ⓐ oxymoron
Ⓑ hyperbole
Ⓒ simile
Ⓓ metaphor

Read the excerpt from My Antonia by Willa Cather, then answer the questions.

"I had a sense of coming home to myself, and of having found out what a little circle man's experience is. For Ántonia and for me, this had been the road of Destiny; had taken us to those early accidents of fortune which predetermined for us all that we can ever be. Now I understood that the same road was to bring us together again. Whatever we had missed, we possessed together the precious, the incommunicable past."

9. What is the meaning of the word "predetermined" as it is used in this context?

Ⓐ an accident or unfortunate circumstance
Ⓑ to be decided in advance
Ⓒ easy to agree upon
Ⓓ something that is flexible or unsettled

10. Which statement best reflects the theme of this passage?

Ⓐ A person cannot escape his or her past; it is always a part of them.
Ⓑ Nature is often a reflection of what a person is feeling inside.
Ⓒ Humans must overcome countless obstacles to realize the true meaning of life.
Ⓓ It is natural to fear "the other", or those who are different from yourself.

DAY 5

CHALLENGE YOURSELF!

✔ Describing Objects and Their Properties
✔ Writing-Language

www.lumoslearning.com/a/dc10-40

See the first page for Signup details

WEEK 8 - ESSAY PROMPT

✔ Exploring Academic Interests

www.lumoslearning.com/a/wc10-w8

See the first page for Signup details

Cracked The SAT! Now What?

College Options

Now it is time to consider college. First, you will want to look at the admission statistics of colleges you would like to attend and see what the average score for admittance is. This will help you to narrow down college options. Do not be afraid to apply if your score is lower than what they suggest; you can make up for that in other areas of the application.

Scholarships and Grants

Scoring high on the SAT allows you to get grants and scholarships to help pay for your education. Some of them you do not even have to apply for. Others have an application process and you will want to be sure you meet the deadline.

Some smaller or private scholarships will ask you to submit letters of recommendation and/or essays, and the recipient will be chosen based on those as well as other factors. Some scholarships are based on personal circumstances or even what your parents do for employment.

Colleges and universities tend to offer scholarships based on merit and test scores. Many of these will be awarded automatically and there is no additional application needed.

The Bottom Line

Now that you have your SAT score, you have to decide if you are happy with it. If you did not score as well as you would like or feel you need to, you can take the test again. If you have a score with which you are happy and confident, use it to move forward with your college or university application process.

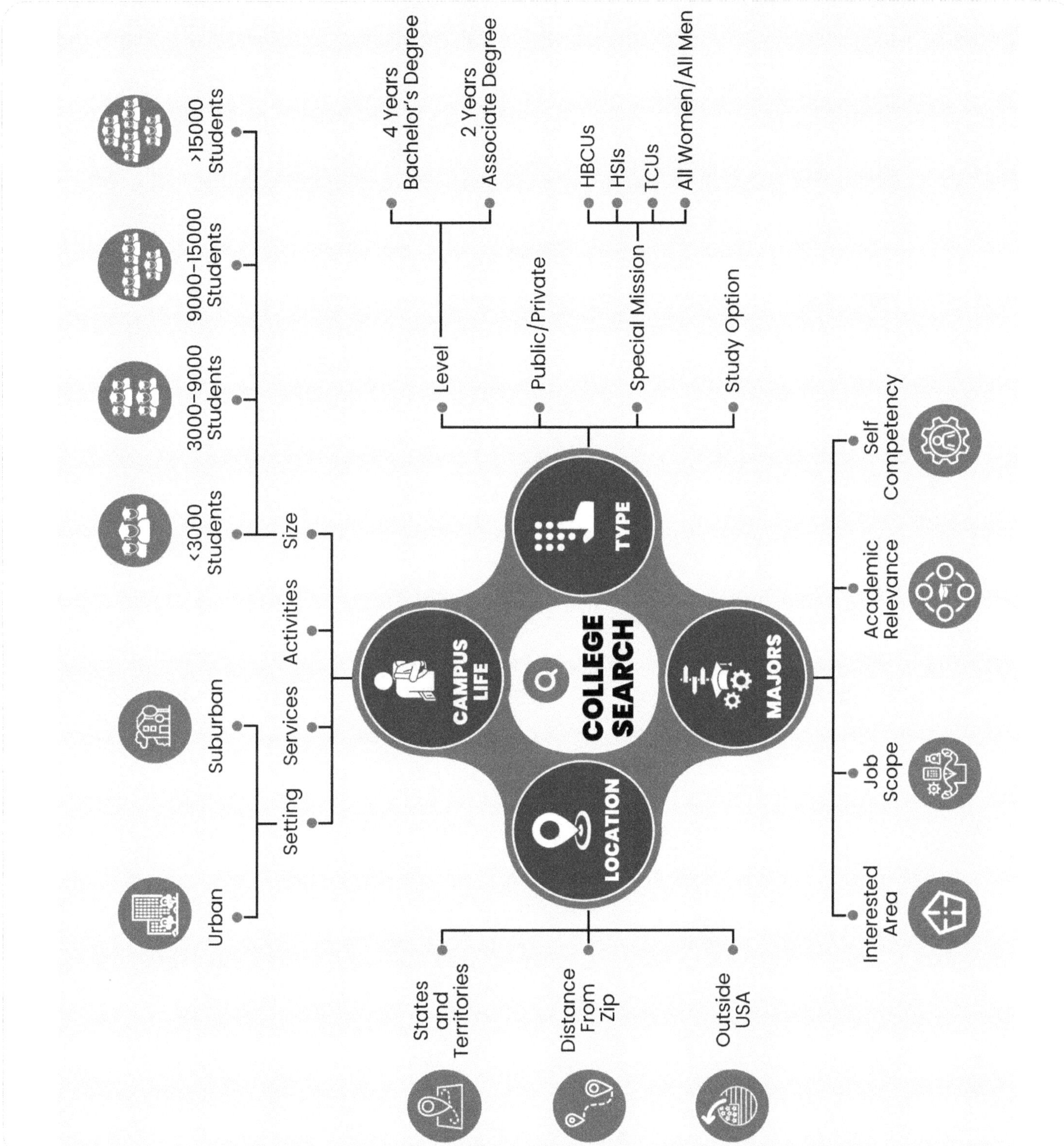

COLLEGE READINESS MIND MAPS

✔ College-Search

www.lumoslearning.com/a/crm8

See the first page for Signup details

WEEK 9
SUMMER PRACTICE

INTERPRET EXPRESSIONS

DAY 1

1. What is the term with the highest degree in the expression $7x^9y-9xy^8+2x^5y^6-5xy$?

Ⓐ $7x^9y$
Ⓑ $-9xy^8$
Ⓒ $2x^5y^6$
Ⓓ $-5xy$

2. What is a factor of the expression $36x^2+12x+24$?

Ⓐ x
Ⓑ 12
Ⓒ 24
Ⓓ 12x

3. The formula $P(1+r)^t$ is used to calculate the amount in an account that earns interest com pounded annually. Here, P is the principal in the account, r is the annual interest rate (as a decimal), and t is the time (in years). What is the interest rate in the expression 500(1.025)?

Ⓐ 1.5%
Ⓑ 2.5%
Ⓒ 1.025%
Ⓓ 25%

4. The formula $P(1+r)^t$ is used to calculate the amount in an account that earns interest compounded annually, when, *P* is the principal in the account, *r* is the annual interest rate(as a decimal), and t is the time(in years). If the expression $1500(1.0355)^9$ was used to calculate the amount of money in an account, for how many years was the money in the account?

Ⓐ 7
Ⓑ 9%
Ⓒ 9
Ⓓ 7%

5. Thomas bought a baseball bat and two gloves for a total of *d* dollars. If *g* represents the cost of one glove, what expression represents the cost of the baseball bat?

Ⓐ d + 2g
Ⓑ 2g - d
Ⓒ d - 2g
Ⓓ $d + \frac{9}{2}$

The Most Dangerous Game

By Richard Connell

Read the passage below and answer the questions that follow.

"OFF THERE to the right--somewhere--is a large island," said Whitney." It's rather a mystery--" "What island is it?" Rainsford asked.

"The old charts call it `Ship-Trap Island,'" Whitney replied." A suggestive name, isn't it? Sailors have a curious dread of the place. I don't know why. Some superstition--" "Can't see it," remarked Rainsford, trying to peer through the dank tropical night that was palpable as it pressed its thick warm blackness in upon the yacht.

"You've good eyes," said Whitney, with a laugh," and I've seen you pick off a moose moving in the brown fall bush at four hundred yards, but even you can't see four miles or so through a moonless Caribbean night."

"Nor four yards," admitted Rainsford. "Ugh! It's like moist black velvet."

"It will be light enough in Rio," promised Whitney. "We should make it in a few days. I hope the jaguar guns have come from Purdey's. We should have some good hunting up the Amazon. Great sport, hunting."

"The best sport in the world," agreed Rainsford.

"For the hunter," amended Whitney. "Not for the jaguar."

"Don't talk rot, Whitney," said Rainsford.

"You're a big-game hunter, not a philosopher. Who cares how a jaguar feels?"
"Perhaps the jaguar does," observed Whitney.

"Bah! They've no understanding."

"Even so, I rather think they understand one thing--fear. The fear of pain and the fear of death."

"Nonsense," laughed Rainsford. "This hot weather is making you soft, Whitney. Be a realist. The world is made up of two classes--the hunters and the huntees. Luckily, you and I are hunters. Do you think we've passed that island yet?"

"I can't tell in the dark. I hope so."
"Why? " asked Rainsford.

"The place has a reputation--a bad one."

"Cannibals?" suggested Rainsford.

"Hardly. Even cannibals wouldn't live in such a God-forsaken place. But it's gotten into sailor lore, somehow. Didn't you notice that the crew's nerves seemed a bit jumpy today?"

"They were a bit strange, now you mention it. Even Captain Nielsen--"

"Yes, even that tough-minded old Swede, who'd go up to the devil himself and ask him for a light. Those fishy blue eyes held a look I never saw there before. All I could get out of him was `This place has an evil name among seafaring men, sir.' Then he said to me, very gravely, `Don't you feel anything?'--as if the air about us was actually poisonous. Now, you mustn't laugh when I tell you this--I did feel something like a sudden chill.

"There was no breeze. The sea was as flat as a plate-glass window. We were drawing near the island then. What I felt was a--a mental chill; a sort of sudden dread."

"Pure imagination," said Rainsford.

"One superstitious sailor can taint the whole ship's company with his fear."

"Maybe. But sometimes I think sailors have an extra sense that tells them when they are in danger. Sometimes I think evil is a tangible thing--with wavelengths, just as sound and light have. An evil place can, so to speak, broadcast vibrations of evil. Anyhow, I'm glad we're getting out of this zone. Well, I think I'll turn in now, Rainsford."

"I'm not sleepy," said Rainsford. "I'm going to smoke another pipe up on the afterdeck."

"Good night, then, Rainsford. See you at breakfast."

"Right. Good night, Whitney."

There was no sound in the night as Rainsford sat there but the muffled throb of the engine that drove the yacht swiftly through the darkness, and the swish and ripple of the wash of the propeller.

Rainsford, reclining in a steamer chair, indolently puffed on his favorite brier. The sensuous drowsiness of the night was on him." It's so dark," he thought, "that I could sleep without closing my eyes; the night would be my eyelids--"

An abrupt sound startled him. Off to the right he heard it, and his ears, expert in such matters, could not be mistaken. Again he heard the sound, and again. Somewhere, off in the blackness, someone had fired a gun three times.

Rainsford sprang up and moved quickly to the rail, mystified.

He strained his eyes in the direction from which the reports had come, but it was like trying to see through a blanket. He leaped upon the rail and balanced himself there, to get greater elevation; his pipe, striking a rope, was knocked from his mouth. He lunged for it; a short, hoarse cry came from his lips as he realized he had reached too far and had lost his balance. The cry was pinched off short as the blood-warm waters of the

Caribbean Sea closed over his head.

He struggled up to the surface and tried to cry out, but the wash from the speeding yacht slapped him in the face and the salt water in his open mouth made him gag and strangle. Desperately he struck out with strong strokes after the receding lights of the yacht, but he stopped before he had swum fifty feet. A certain cool headedness had come to him; it was not the first time he had been in a tight place. There was a chance that his cries could be heard by someone aboard the yacht, but that chance was slender and grew more slender as the yacht raced on. He wrestled himself out of his clothes and shouted with all his power. The lights of the yacht became faint and ever-vanishing fireflies; then they were blotted out entirely by the night. Rainsford remembered the shots. They had come from the right, and doggedly he swam in that direction, swimming with slow, deliberate strokes, conserving his strength. For a seemingly endless time he fought the sea. He began to count his strokes; he could do possibly a hundred more and then Rainsford heard a sound. It came out of the darkness, a high screaming sound, the sound of an animal in an extremity of anguish and terror.

He did not recognize the animal that made the sound; he did not try to; with fresh vitality he swam toward the sound. He heard it again; then it was cut short by another noise, crisp, staccato.

"Pistol shot," muttered Rainsford, swimming on.

Ten minutes of determined effort brought another sound to his ears the most welcome he had ever heard the muttering and growling of the sea breaking on a rocky shore. He was almost on the rocks before he saw them; on a night less calm he would have been shattered against them.

With his remaining strength he dragged himself from the swirling waters. Jagged crags appeared to jut up into the opaqueness; he forced himself upward, hand over hand. Gasping, his hands raw, he reached a flat place at the top. Dense jungle came down to the very edge of the cliffs. What perils that tangle of trees and underbrush might hold for him did not concern Rainsford just then. All he knew was that he was safe from his enemy, the sea, and that utter weariness was on him. He flung himself down at the jungle edge and tumbled headlong into the deepest sleep of his life.

When he opened his eyes he knew from the position of the sun that it was late in the afternoon. Sleep had given him new vigor; a sharp hunger was picking at him. He looked about him, almost cheerfully.

"Where there are pistol shots, there are men. Where there are men, there is food," he thought. But what kind of men, he wondered, in so forbidding a place? An unbroken front of snarled and ragged jungle fringed the shore.

He saw no sign of a trail through the closely knit web of weeds and trees; it was easier to go along the shore, and Rainsford floundered along by the water. Not far from where he landed, he stopped.

Some wounded thing by the evidence, a large animal had thrashed about in the underbrush; the jungle weeds were crushed down and the moss was lacerated; one patch of weeds was stained crimson. A small, glittering object not far away caught Rainsford's eye and he picked it up. It was an empty cartridge.

"A twenty-two," he remarked. "That's odd. It must have been a fairly large animal too. The hunter had his nerve with him to tackle it with a light gun. It's clear that the brute put up a fight. I suppose the first three shots I heard was when the hunter flushed his quarry and wounded it. The last shot was when he trailed it here and finished it." He examined the ground closely and found what he had hoped to find, the print of hunting boots. They pointed along the cliff in the direction he had been going. Eagerly he hurried along, now slipping on a rotten log or a loose stone, but making headway; night was beginning to settle down on the island.

Bleak darkness was blacking out the sea and jungle when Rainsford sighted the lights. He came upon them as he turned a crook in the coast line; and his first thought was that he had come upon a village, for there were many lights. But as he forged along he saw to his great astonishment that all the lights were in one enormous building a lofty structure with pointed towers plunging upward into the gloom. His eyes made out the shadowy outlines of a palatial chateau; it was set on a high bluff, and on three sides of it cliffs dived down to where the sea licked greedy lips in the shadows.

"Mirage," thought Rainsford. But it was no mirage, he found, when he opened the tall spiked iron gate. The stone steps were real enough; the massive door with a leering gargoyle for a knocker was real enough; yet above it all hung an air of unreality.

He lifted the knocker, and it creaked up stiffly, as if it had never before been used. He let it fall, and it startled him with its booming loudness. He thought he heard steps within; the door remained closed. Again Rainsford lifted the heavy knocker, and let it fall. The door then opened as suddenly as if it were on a spring and Rainsford stood blinking in the river of glaring gold light that poured out. The first thing Rainsford's eyes discerned was the largest man Rainsford had ever seen, a gigantic creature, solidly made and black-bearded to the waist. In his hand the man held a long-barreled revolver, and he was pointing it straight at Rainsford's heart.

Out of the snarl of beard two small eyes regarded Rainsford.

"Don't be alarmed," said Rainsford, with a smile which he hoped was disarming. "I'm no robber. I fell off a yacht. My name is Sanger Rainsford of New York City."

The menacing look in the eyes did not change. The revolver pointing as rigidly as if the giant were a statue.

He gave no sign that he understood Rainsford's words, or that he had even heard them. He was dressed in uniform a black uniform trimmed with gray astrakhan.

"I'm Sanger Rainsford of New York," Rainsford began again. "I fell off a yacht. I am hungry."

The man's only answer was to raise with his thumb the hammer of his revolver. Then Rainsford saw the man's free hand go to his forehead in a military salute, and he saw him click his heels together and stand at attention. Another man was coming down the broad marble steps, an erect, slender man in evening clothes.
He advanced to Rainsford and held out his hand.

In a cultivated voice marked by a slight accent that gave it added precision and deliberateness, he said, "It is a very great pleasure and honor to welcome Mr. Sanger Rainsford, the celebrated hunter, to my home."

In a cultivated voice marked by a slight accent that gave it added precision and deliberateness, he said, "It is a very great pleasure and honor to welcome Mr. Sanger Rainsford, the celebrated hunter, to my home."

Automatically Rainsford shook the man's hand.

"I've read your book about hunting snow leopards in Tibet, you see," explained the man. "I am General Zaroff."

Rainsford's first impression was that the man was singularly handsome; his second was that there was an original, almost bizarre quality about the general's face. He was a tall man past middle age, for his hair was a vivid white; but his thick eyebrows and pointed military mustache were as black as the night from which Rainsford had come. His eyes, too, were black and very bright. He had high cheekbones, a sharpcut nose, a spare, dark face the face of a man used to giving orders, the face of an aristocrat. Turning to the giant in uniform, the general made a sign. The giant put away his pistol, saluted, withdrew.

"Ivan is an incredibly strong fellow," remarked the general, "but he has the misfortune to be deaf and dumb. A simple fellow, but, I'm afraid, like all his race, a bit of a savage."

"Is he Russian?"

"He is a Cossack," said the general, and his smile showed red lips and pointed teeth. "So am I."

"Come," he said, "we shouldn't be chatting here. We can talk later. Now you want clothes, food, rest. You shall have them. This is a most-restful spot."

Ivan had reappeared, and the general spoke to him with lips that moved but gave forth no sound.

"Follow Ivan, if you please, Mr. Rainsford," said the general. "I was about to have my dinner when you came. I'll wait for you. You'll find that my clothes will fit you, I think."

It was to a huge, beam-ceilinged bedroom with a canopied bed big enough for six men that Rainsford followed the silent giant. Ivan laid out an evening suit, and Rainsford, as he put it on, noticed that it came from a London tailor who ordinarily cut and sewed for none below the rank of duke.

The dining room to which Ivan conducted him was in many ways remarkable. There was a medieval magnificence about it; it suggested a baronial hall of feudal times with its oaken panels, its high ceiling, its vast refectory tables where twoscore men could sit down to eat. About the hall were mounted heads of many animals, lions, tigers, elephants, moose, bears; larger or more perfect specimens Rainsford had never seen.At the great table the general was sitting, alone.

"You'll have a cocktail, Mr. Rainsford," he suggested. The cocktail was surpassingly good; and, Rainsford noted, the table appointments were of the finest--the linen, the crystal, the silver, the china.
They were eating borsch, the rich, red soup with whipped cream so dear to Russian palates.

Half apologetically General Zaroff said, "We do our best to preserve the amenities of civilization here. Please forgive any lapses. We are well off the beaten track, you know. Do you think the champagne has suffered from its long ocean trip?"

"Not in the least," declared Rainsford. He was finding the general a most thoughtful and affable host, a true cosmopolite. But there was one small trait of the general's that made Rainsford uncomfortable. Whenever he looked up from his plate he found the general studying him, appraising him narrowly.

"Perhaps," said General Zaroff, "you were surprised that I recognized your name. You see, I read all books on hunting published in English, French, and Russian. I have but one passion in my life, Mr. Rainsford, and it is the hunt."

"You have some wonderful heads here," said Rainsford as he ate a particularly well-cooked filet mignon."

That Cape buffalo is the largest I ever saw."

"Oh, that fellow. Yes, he was a monster."

"Did he charge you?"

"Hurled me against a tree," said the general. "Fractured my skull. But I got the brute."

"I've always thought," said Rains{ord, "that the Cape buffalo is the most dangerous of all big game."
For a moment the general did not reply; he was smiling his curious red-lipped smile. Then he said slowly, "No. You are wrong, sir. The Cape buffalo is not the most dangerous big game." He sipped his wine. "Here in my preserve on this island," he said in the same slow tone, "I hunt more dangerous game."

"Hurled me against a tree," said the general. "Fractured my skull. But I got the brute."

"I've always thought," said Rainsford, "that the Cape buffalo is the most dangerous of all big game."

For a moment the general did not reply; he was smiling his curious red-lipped smile. Then he said slowly, "No. You are wrong, sir. The Cape buffalo is not the most dangerous big game." He sipped his wine. "Here in my preserve on this island," he said in the same slow tone, "I hunt more dangerous game."

6. Select the quote that best reflects the story's theme that "Human life is meant to be protected rather than hunted."

Ⓐ "Life is for the strong, to be lived by the strong, and, if needs be, taken by the strong."
Ⓑ "I refuse to believe that so modern and civilized a young man as you seem to be harbors romantic ideas about the value of human life."
Ⓒ "Hunting? Great Guns, General Zaroff, what you speak of is murder."
Ⓓ "Every day I hunt, and I never grow bored now, for I have a quarry with which I can match my wits."

Read the following passage from "The Most Dangerous Game" by Richard Connell.

"So," continued the general, "I asked myself why the hunt no longer fascinated me. You are much younger than I am, Mr. Rainsford, and have not hunted as much, but you perhaps can guess the answer."

"What was it?"

"Simply this: hunting had ceased to be what you call `a sporting proposition.' It had become too easy. I always got my quarry. Always. There is no greater bore than perfection."

7. The general equates unlimited success to

Ⓐ sport
Ⓑ estimation
Ⓒ ease
Ⓓ boredom

Read the following excerpt and answer the question that follows.

In a cultivated voice marked by a slight accent that gave it added precision and deliberateness, he said, "It is a very great pleasure and honor to welcome Mr. Sanger Rainsford, the celebrated hunter, to my home."

Automatically Rainsford shook the man's hand.

"I've read your book about hunting snow leopards in Tibet, you see," explained the man. "I am General Zaroff."

8. Based on the excerpt above, General Zaroff's attitude toward Rainsford seems to be:

Ⓐ pity
Ⓑ admiration
Ⓒ disgust
Ⓓ fear

9. Which of the following lines does NOT contain a simile?

Ⓐ "He strained his eyes in the direction from which the reports had come, but it was like trying to see through a blanket."

Ⓑ "There was no sound in the night as Rainsford sat there but the muffled throb of the engine that drove the yacht swiftly through the darkness, and the swish and ripple of the wash of the propeller."

Ⓒ "One does not expect nowadays to find a young man of the educated class, even in America, with such a naive, and, if I may say so, mid-Victorian point of view. It's like finding a snuffbox in a limousine."

Ⓓ "They indicate a channel," he said, "where there's none; giant rocks with razor edges crouch like a sea monster with wide-open jaws."

10. The below text contains what type of figurative language?

The general was playing with him! The general was saving him for another day's sport! The Cossack was the cat; he (Rainsford) was the mouse.

Ⓐ simile
Ⓑ metaphor
Ⓒ personification
Ⓓ hyperbole

DAY 1

CHALLENGE YOURSELF!

✓ Interpret Expressions
✓ Reading-Literature

www.lumoslearning.com/a/dc10-41

See the first page for Signup details

1. Rewrite the expression $y^{\frac{4}{5}}$ as an expression with rational exponents, using the properties of exponents.

Ⓐ $\sqrt[5]{y^4}$
Ⓑ $\sqrt[4]{y^5}$
Ⓒ $4\sqrt[5]{y}$
Ⓓ $\frac{\sqrt{y^4}}{5}$

2. Rewrite the expression $(xyz)^{\frac{3}{8}}$ as an expression with rational exponents, using the properties of exponents.

Ⓐ $\sqrt[8]{(xyz)^3}$
Ⓑ $\sqrt[3]{(xyz)^8}$
Ⓒ $(xyz)^{\frac{3}{8}}$
Ⓓ $3(xyz)^{\frac{1}{8}}$

3. Rewrite the radical expression $\sqrt[7]{x^5y^6}$ as an expression with rational exponents, using the properties of exponents.

Ⓐ $x^{\frac{7}{5}}y^{\frac{7}{6}}$
Ⓑ $(x^5y^6)^7$
Ⓒ $(x^{\frac{1}{5}}y^{\frac{1}{6}})^7$
Ⓓ $x^{\frac{5}{7}}y^{\frac{6}{7}}$

4. Rewrite the radical expression $\sqrt[3]{x^3y^6z^9}$ as an expression with rational exponents, using the properties of exponents.

Ⓐ xy^2z^3
Ⓑ $x^{\frac{3}{3}}y^{\frac{3}{6}}z^{\frac{3}{9}}$
Ⓒ $x^{\frac{1}{2}}y^{\frac{1}{3}}z$
Ⓓ $(x^3y^6z^9)^3$

5. Rewrite the radical expression $\sqrt[4]{(abc)^9}$ as an expression with rational exponents, using the properties of exponents.

Ⓐ $(abc)^{\frac{9}{4}}$
Ⓑ $(abc)^{\frac{4}{9}}$
Ⓒ $(abc)^{36}$
Ⓓ $(abc)^5$

6. Which of the following is a part of a perfect paragraph?

Ⓐ Topic Sentence
Ⓑ Supporting Sentences
Ⓒ Clincher
Ⓓ All of the above
Ⓔ None of the above

7. In a traditional 5-paragraph argumentative essay, how many points should you discuss in your thesis?

Ⓐ One
Ⓑ Seven
Ⓒ Three
Ⓓ Two

8. Which paragraph includes the thesis in a 5 paragraph essay?

Ⓐ Supporting Paragraph #1
Ⓑ Conclusion
Ⓒ Introduction
Ⓓ None of the above

9. What is a topic sentence?

Ⓐ A sentence that has to link back to your thesis.
Ⓑ The first sentence of a paragraph.
Ⓒ A sentence that tells what your paragraph will be about.
Ⓓ All of the above.
Ⓔ None of the above.

10. Which sentence is the best thesis statement?

Ⓐ Even though McDonalds seems to be the food of choice for children, many families are rejecting the pressure to feed children there and are choosing higher-quality food.
Ⓑ I am going to explain why children should eat better food than McDonalds.
Ⓒ McDonalds is bad.
Ⓓ None of the above.

DAY 2

CHALLENGE YOURSELF!

✓ Simplifying Expressions with Rational Exponents
✓ Writing

www.lumoslearning.com/a/dc10-42

See the first page for Signup details

1. A circle has a central angle that measures 23^0. Find the measure of the angle in radians. Round your answer to the nearest hundredth.

Ⓐ .40
Ⓑ .13
Ⓒ 1317.80
Ⓓ .20

2. A circle has a central angle of 240°. Find the measure of the angle in radians. Give the exact value of the angle (leave the measure in terms of π).

Ⓐ $\frac{3\pi}{4}$
Ⓑ $\frac{4\pi}{3}$
Ⓒ $\frac{5\pi}{6}$
Ⓓ $\frac{6\pi}{5}$

3 The exact value of the central angle of a circle is $\frac{3\pi}{2}$. Find the measure of the central angle of the circle in degrees.

Ⓐ 60°
Ⓑ 240°
Ⓒ 120°
Ⓓ 270°

4. Find the degree and radian measures of the angle in standard position by rotating the terminal side by $\frac{1}{3}$ of a circle.

Ⓐ $120^0, \frac{2\pi}{3}$
Ⓑ $120^0, \frac{3\pi}{2}$
Ⓒ $60^0, \frac{\pi}{3}$
Ⓓ $60^0, \frac{\pi}{6}$

5. What is the length, in inches, of an arc subtended by an angle of 36^0 on a circle with a radius of 15 inches?

Ⓐ 540
Ⓑ 36π
Ⓒ 36
Ⓓ 3π

Scoundrel History and Utopian Method

Read the passage below and answer the question that follows.

Le Libertaire, iii, 1895, presumed Louise Michel

Who among us does not feel the shadow of fear cast by the cowardly laws of these past years? The Scoundrel Laws terrorize not only those who might commit violence but anyone who associates with them. They reward those who denounce their brothers and sisters, sowing distrust and ill-will. They freeze our hearts and our tongues, by punishing with prison anyone who provokes, praises, or merely seeks to understand those mad acts to which an insane society has driven a few poor souls.

Perhaps even these words, here, are enough to summon our new inquisitors.

If so, I say, let them come. I know their jail cells; their guards are my comrades and friends. Scoundrel laws, like the scoundrels who created them, must one day lose their power. It is a law of justice and nature.

We who know the future, who see with certainty like a memory ahead of us the society of freedom, equality, brotherhood, and sisterhood, we learn such laws from the past, even, at times, from the bourgeois chroniclers behind the walls of the Sorbonne. At the very least, from them, we may learn what new tricks they employ.

Thus I recently read the three volumes of the historian whom they call "great", Hippolyte Taine, to learn what lies were being passed off as the official past. Taine sat in a special Chair, for the History of the Revolution – a monument propped on the grave of the past like a tombstone, to guarantee that such an event will never happen again.

This historian is now himself buried, covered in praise. I come not to desecrate his memory, but merely, after so many éloges by that mindless thinkers-by-the-hour, to restore it to its true and laughable scale.

Taine tells us that history is an obscure knowledge of a distant past and that we have nothing to expect from the future. The Revolution, this decisive overflowing of the desire for justice and a better life, is denounced as a deception, a struggle for power among upstarts, a monstrosity never to be repeated.

Among all Taine's half-truths, fantasies, and idiocies, none is so great as his pretension to science – in which history is always a balance between race, milieu, and moment. He transforms inspired and courageous men and women into dupes of nature and of their fellows; by this, he hopes to complete whatever bestialization the rulers of this society have not yet wrought.

He has created the equivalent in letters of those villainous acts of legislation. He has written l'Histoire scélérate, Scoundrel History. It shuts people's mouths and severs their connection to the dreams, sweat, and aspirations of those who struggled before us. Scoundrel History insists on the difference between now and then, the arbitrariness of the new, the fatalism of birth, of rocks, vegetation, and rivers. In the name of science, he lashes those who embraced a world more vast than his vanity.

Were these his only crimes, I would happily cast his miserable books aside – or in a more generous spirit, wrap fish in them, so that in some small way they might serve life. But I am moved to take up my pen, finally, by his third volume, which blesses the current power as good, just, and in any case inevitable. What spurs me to write is a single citation, a unique if profound mistake.

6. What argument is the author making in this passage?

Ⓐ We can't know our future.
Ⓑ Knowing our history can help us to identify our future.
Ⓒ The struggle for power hinders our view of the future.
Ⓓ The system is built against us as a society.

Read the following excerpt from "Simple Gifts", by Joseph Bracket, Jr. and answer the following question.

'Tis the gift to be loved and that love to return,
'Tis the gift to be taught and a richer gift to learn,
And when we expect of others what we try to live each day,
Then we'll all live together and we'll all learn to say,
Tis the gift to be simple, 'tis the gift to be free,
'Tis the gift to come down where we ought to be,
And when we find ourselves in the place just right,
'Twill be in the valley of love and delight.
When true simplicity is gained,
To bow and to bend we shan't be ashamed,
To turn, turn will be our delight,
Till by turning, turning we come round right.

7. Which line of the song BEST supports the author's message?

Ⓐ And when we expect of others what we try to live each day,
Ⓑ Till by turning, turning we come round right.
Ⓒ Then we'll all live together and we'll all learn to say
Ⓓ When true simplicity is gained,

Address the Nation from the Challenger

- speech from President Ronald Reagan

Read the passage below and answer the questions that follow.

Ladies and Gentlemen, I'd planned to speak to you tonight to report on the state of the Union, but the events of earlier today have led me to change those plans. Today is a day for mourning and remembering. Nancy and I are pained to the core by the tragedy of the shuttle Challenger. We know we share this pain with all of the people of our country. This is truly a national loss.

Nineteen years ago, almost to the day, we lost three astronauts in a terrible accident on the ground. But, we've never lost an astronaut in flight; we've never had a tragedy like this. And perhaps we've forgotten the courage it took for the crew of the shuttle; but they, the Challenger Seven, were aware of the dangers, but overcame them and did their jobs brilliantly.We mourn seven heroes: Michael Smith, Dick Scobee, Judith Resnik, Ronald McNair, Ellison Onizuka, Gregory Jarvis, and Christa McAuliffe. We mourn their loss as a nation together.

For the families of the seven, we cannot bear, as you do, the full impact of this tragedy. But we feel the loss, and we're thinking about you so very much. Your loved ones were daring and brave, and they had that special grace, that special spirit that says, 'Give me a challenge and I'll meet it with joy.' They had a hunger to explore the universe and discover its truths. They wished to serve, and they did. They served all of us.

We've grown used to wonders in this century. It's hard to dazzle us. But for twenty-five years the United States space program has been doing just that. We've grown used to the idea of space, and perhaps we forget that we've only just begun. We're still pioneers. They, the members of the Challenger crew, were pioneers.

And I want to say something to the school children of America who were watching the live coverage of the shuttle's takeoff. I know it is hard to understand, but sometimes painful things like this happen. It's all part of the process of exploration and discovery. It's all part of taking a chance and expanding man's horizons. The future doesn't belong to the fainthearted; it belongs to the brave. The Challenger crew was pulling us into the future, and we'll continue to follow them.

I've always had great faith in and respect for our space program, and what happened today does nothing to diminish it. We don't hide our space program. We don't keep secrets and cover things up. We do it all up front and in public. That's the way freedom is, and we wouldn't change it for a minute. We'll continue our quest in space. There will be more shuttle flights and more shuttle crews and, yes, more volunteers, more civilians, more teachers in space. Nothing ends here; our hopes and our journeys continue. I want to add that I wish I could talk to every man and woman who works for NASA or who worked on this mission and tell them: "Your dedication and professionalism have moved and impressed us for decades. And we know of your anguish. We share it."

There's a coincidence today. On this day 390 years ago, the great explorer Sir Francis Drake died aboard ship off the coast of Panama. In his lifetime the great frontiers were the oceans, and a historian later said, 'He lived by the sea, died on it, and was buried in it.' Well, today we can say of the Challenger crew: Their dedication was, like Drake's, complete.

The crew of the space shuttle Challenger

honoured us by the manner in which they lived their lives. We will never forget them, nor the last time we saw them this morning as they prepared for the journey and waved goodbye and 'slipped the surly bonds of earth' to 'touch the face of God.'

Thank you.

8. What is the OVERALL claim that is being made in this speech?

Ⓐ To unite as a country
Ⓑ To remember those who have passed
Ⓒ The great work that NASA has done
Ⓓ Understanding why things happen

9. These words are very important in an argument because they allow you to move between ideas helping you to connect, compare, and contrast your sentences and paragraphs:

Ⓐ transitions
Ⓑ connectives
Ⓒ metaphors
Ⓓ paragraphs

10. Determine whether the following statement is True or False.

In an argument, opinions must be proven with facts, expert testimonies, and other evidence to be considered valid.

Ⓐ True
Ⓑ False

1. Given that ΔABC ≅ ΔEFG, E=30⁰, B=124⁰, and what is the measure of C?

Ⓐ 26^0
Ⓑ 30^0
Ⓒ 56^0
Ⓓ 150^0

2. If ΔBCD ≅ ΔMNO and the vertices of ΔBCD are located at B(-2, -2), C(4, -2), D(4, 6), determine the length of segment NO.

Ⓐ 6
Ⓑ 10
Ⓒ 11.66
Ⓓ 4

3. Given that ΔXYZ≈ΔABC, XY=AB and YZ=BC, which additional statement must also be true?

Ⓐ BA=YX
Ⓑ BC=BC
Ⓒ ZX=CA
Ⓓ ZX=BA

4. The congruence markings on one of the triangles in the figure below are incorrect. Exactly two of the three triangles in the figure are congruent. Which congruency statement correctly states which two triangles are congruent?

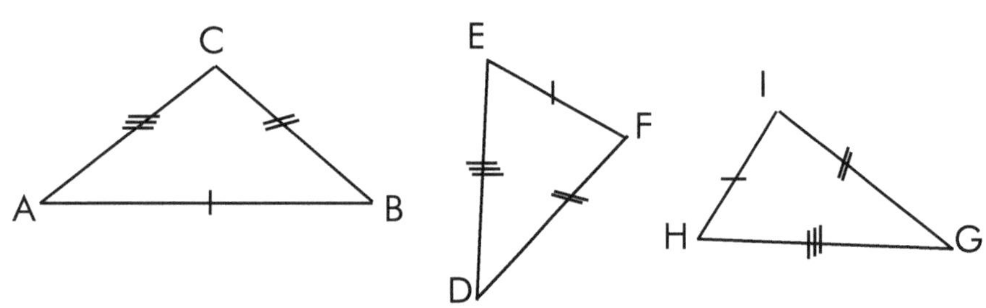

Ⓐ ΔABC ≅ ΔHGI
Ⓑ ΔDEF ≅ ΔABC
Ⓒ ΔDEF ≅ ΔGHI
Ⓓ ΔDEF ≅ ΔIGH

5. The congruency markings on one of the triangles in the figure below are incorrect. Two of the three triangles in the figure are congruent. Which congruency statement correctly says which two triangles are congruent?

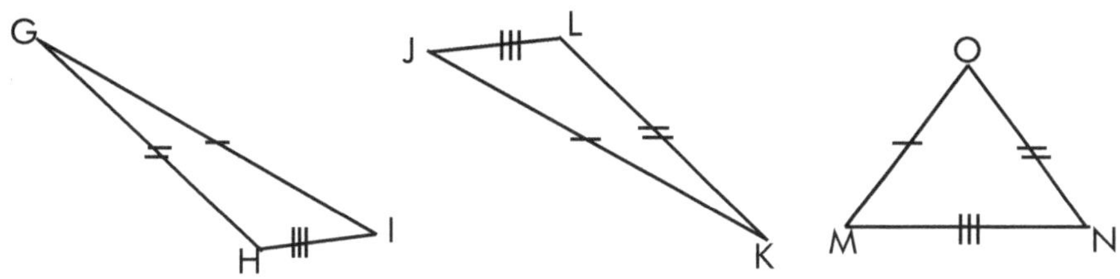

Ⓐ ΔGHI ≅ ΔKLJ
Ⓑ ΔMON ≅ ΔHGI
Ⓒ ΔGHI ≅ ΔKJL
Ⓓ ΔJKL ≅ ΔMNO

6. Choose the academic term that best completes the sentence. "I can ___ from the clothes and hairstyles in this photograph that it was taken during the 1980s."

Ⓐ infer
Ⓑ analyze
Ⓒ evaluate
Ⓓ summarize

7. Which of the following sentences is missing a comma?

Ⓐ When we finished we drove home.
Ⓑ Susan and Craig rode together to school.
Ⓒ We went home and slept.
Ⓓ Jack ran for a mile.

8. When an author exploits the different possible meanings of a word to make a joke, this is referred to as a

Ⓐ hyperbole
Ⓑ pun
Ⓒ oxymoron
Ⓓ onomatopoeia

9. Read the sentence. Based on the context, determine the meaning of the underlined word. "Having left a job where the employees were mostly <u>homogenous</u>, Cindy was glad to work in an office with more fellow women employees."

Ⓐ homebodies
Ⓑ of the same kind
Ⓒ sexist
Ⓓ intelligent

10. Read the sentence. Based on the context, determine the meaning of the underlined word. The accountant expected to see equality between the client's records and his, but he instead found a <u>disparity</u>.

Ⓐ similarity
Ⓑ inequality
Ⓒ growth
Ⓓ decrease

DAY 4

CHALLENGE YOURSELF!

✔ Triangle Congruence Statements
✔ Language

www.lumoslearning.com/a/dc10-44

See the first page for Signup details

1. A globe that models the shape of the earth has a diameter of 80 mm. What is the volume of the globe, rounded to the nearest cubic millimeter (mm^3)?

Ⓐ 268,082 mm^3
Ⓑ 2,144,661 mm^3
Ⓒ 20,106 mm^3
Ⓓ 5,027 mm^3

2. What is the area of a square with a perimeter of 4 feet?

Ⓐ 1 ft^2
Ⓑ 1 ft^2
Ⓒ 2 ft^2
Ⓓ 4 ft^2

3. Ron is building a cylindrical water tank. The circumference of the tank is 6.28 meters. If Ron builds the tank 5 meters tall, what is the volume of water it will hold when completely full? Estimate π=3.14

Ⓐ 15 $meters^3$
Ⓑ 14.7 $meters^3$
Ⓒ 14 $meters^3$
Ⓓ 15.7 $meters^3$

4. Cynthia is attempting to find the volume of a tree trunk. She knows the circumference is 5π feet. If the trunk is about 10 feet tall, what is the best estimate for the volume of the trunk?

Ⓐ 2.5 $feet^3$
Ⓑ 62.5π $feet^3$
Ⓒ 6.2π $feet^3$
Ⓓ 25 $feet^3$

5. John is painting a fence around a rectangular yard. The house forms one side of the fencing which is 30 feet in length and will not be painted. The yard is 30 feet long and 20 feet wide. If the fence is 3 feet high and a gallon of paint covers 50 ft^2, how many gallons should John purchase.

Ⓐ 3 gallons
Ⓑ 4 gallons
Ⓒ 5 gallons
Ⓓ Cannot be found from the information given

John F Kennedy's inaugural address in 1961

Read the passage below and answer the questions that follow.

Vice President Johnson, Mr. Speaker, Mr. Chief Justice, President Eisenhower, Vice President Nixon, President Truman, Reverend Clergy, fellow citizens:

We observe today not a victory of party but a celebration of freedom--symbolizing an end as well as a beginning--signifying renewal as well as change. For I have sworn before you and Almighty God the same solemn oath our our forebears prescribed nearly a century and three-quarters ago.

The world is very different now. For man holds in his mortal hands the power to abolish all forms of human poverty and all forms of human life. And yet the same revolutionary beliefs for which our forebears fought are still at issue around the globe--the belief that the rights of man come not from the generosity of the state but from the hand of God.

We dare not forget today that we are the heirs of that first revolution. Let the word go forth from this time and place, to friend and foe alike, that the torch has been passed to a new generation of Americans--born in this century, tempered by war, disciplined by a hard and bitter peace, proud of our ancient heritage--and unwilling to witness or permit the slow undoing of those human rights to which this nation has always been committed, and to which we are committed today at home and around the world.

Let every nation know, whether it wishes us well or ill, that we shall pay any price, bear any burden, meet any hardship, support any friend, oppose any foe to assure the survival and the success of liberty.

This much we pledge--and more.

To those old allies whose cultural and spiritual origins we share, we pledge the loyalty of faithful friends. United there is little we cannot do in a host of cooperative ventures. Divided there is little we can do--for we dare not meet a powerful challenge at odds and split asunder.

To those new states whom we welcome to the ranks of the free, we pledge our word that one form of colonial control shall not have passed away merely to be replaced by a far more iron tyranny. We shall not always expect to find them supporting our view. But we shall always hope to find them strongly supporting their own freedom--and to remember that, in the past, those who foolishly sought power by riding the back of the tiger ended up inside.

To those people in the huts and villages of half the globe struggling to break the bonds of mass misery, we pledge our best efforts to help them help themselves, for whatever period is required--not because the communists may be doing it, not because we seek their votes, but because it is right. If a free society cannot help the many who are poor, it cannot save the few who are rich.

To our sister republics south of our border, we offer a special pledge--to convert our good words into good deeds--in a new alliance for progress--to assist free men and free governments in casting off the chains of poverty. But this peaceful revolution of hope cannot become the prey of hostile powers.

Let all our neighbors know that we shall join with them to oppose aggression or subversion anywhere in the Americas. And let every other power know that this Hemisphere intends to remain the master of its own house.

To that world assembly of sovereign states, the United Nations, our last best hope in an age where the instruments of war have far outpaced the instruments of peace, we renew our pledge of support--to prevent it from becoming merely a forum for invective--to strengthen its shield of the new and the weak--and to enlarge the area in which its writ may run.

Finally, to those nations who would make themselves our adversary, we offer not a pledge but a request: that both sides begin anew the quest for peace, before the dark powers of destruction unleashed by science engulf all humanity in planned or accidental self-destruction.

We dare not tempt them with weakness. For only when our arms are sufficient beyond doubt can we be certain beyond doubt that they will never be employed.

But neither can two great and powerful groups of nations take comfort from our present course--both sides overburdened by the cost of modern weapons, both rightly alarmed by the steady spread of the deadly atom, yet both racing to alter that uncertain balance of terror that stays the hand of mankind's final war.

So let us begin anew--remembering on both sides that civility is not a sign of weakness, and sincerity is always subject to proof. Let us never negotiate out of fear. But let us never fear to negotiate.

Let both sides explore what problems unite us instead of belaboring those problems which divide us.

Let both sides, for the first time, formulate serious and precise proposals for the inspection and control of arms--and bring the absolute power to destroy other nations under the absolute control of all nations.

Let both sides, for the first time, formulate serious and precise proposals for the inspection and control of arms--and bring the absolute power to destroy other nations under the absolute control of all nations.

Let both sides seek to invoke the wonders of science instead of its terrors. Together let us explore the stars, conquer the deserts, eradicate disease, tap the ocean depths and encourage the arts and commerce.

Let both sides unite to heed in all corners of the earth the command of Isaiah--to "undo the heavy burdens . . . (and) let the oppressed go free."

And if a beachhead of cooperation may push back the jungle of suspicion, let both sides join in creating a new endeavor, not a new balance of power, but a new world of law, where the strong are just and the weak secure and the peace preserved.

All this will not be finished in the first one hundred days. Nor will it be finished in the first one thousand days, nor in the life of this Administration, nor even perhaps in our lifetime on this planet. But let us begin.

In your hands, my fellow citizens, more than mine, will rest the final success or failure of our course. Since this country was founded, each generation of Americans has been summoned to give testimony to its national loyalty. The graves of young Americans who answered the call to service surround the globe.

Now the trumpet summons us again--not as a call to bear arms, though arms we need--not as a call to battle, though embattled we are-- but a call to bear the burden of a long twilight struggle, year in and year out, "rejoicing in hope, patient in tribulation"--a struggle against the common enemies of man: tyranny, poverty, disease and war itself.

Can we forge against these enemies a grand and global alliance, North and South, East and West, that can assure a more fruitful life for all mankind? Will you join in that historic effort?

In the long history of the world, only a few generations have been granted the role of defending freedom in its hour of maximum danger. I do not shrink from this responsibility--I welcome it. I do not believe that any of us would exchange places with any other people or any other generation. The energy, the faith, the devotion which we bring to this endeavor will light our country and all who serve it--and the glow from that fire can truly light the world.

And so, my fellow Americans: ask not what your country can do for you--ask what you can do for your country.

My fellow citizens of the world: ask not what America will do for you, but what together we can do for the freedom of man.

Finally, whether you are citizens of America or citizens of the world, ask of us here the same high standards of strength and sacrifice which we ask of you. With a good conscience our only sure reward, with history the final judge of our deeds, let us go forth to lead the land we love, asking His blessing and His help, but knowing that here on earth God's work must truly be our own.

6. The purpose of this speech is to:

- Ⓐ To inform the citizens of America about his intentions as a leader and how he intends to work for the progress of the country.
- Ⓑ Emphasize his anger at the war and its effects.
- Ⓒ Question the countrymen about their role and duties.
- Ⓓ None of the above.

7. The theme(s) of Kennedy's speech is:

- Ⓐ Strength
- Ⓑ Compassion
- Ⓒ Hope and belief
- Ⓓ All of the above

8. When accessing online text it is important to

- Ⓐ Evaluate the sources
- Ⓑ Check the validity of what is being presented
- Ⓒ Ignore fake news sources
- Ⓓ All of the above

9. The structure of a short story tends to follow the following plot diagram:

Ⓐ Exposition, rising action, climax, falling action, resolution
Ⓑ Rising action, resolution, exposition, climax, falling action
Ⓒ Character, rising action, exposition, setting, falling action, conflict
Ⓓ Exposition, character, falling action, rising action, theme

10. Which of the following sentences about summarizing is true?

Ⓐ Summarizing teaches the students to isolate what is important.
Ⓑ Summarizing is retelling.
Ⓒ Summarizing helps the student to demonstrate an understanding of the text.
Ⓓ All of the above.

DAY 5

CHALLENGE YOURSELF!

✔ Describing Objects and Their Properties
✔ Reading-Writing-Language

www.lumoslearning.com/a/dc10-45

See the first page for Signup details

WEEK 9 - ESSAY PROMPT

✔ Learning from Conversations

www.lumoslearning.com/a/wc10-w9

See the first page for Signup details

How to Search Colleges Based on Your SAT/ACT Score?

College Search Tools

Once you have your SAT/ACT scores, you can put them into an online college search tool. Three of the most effective are Big Future, College View, and College Data. A good college search tool will have many nuanced filters, rather than just a generic list of courses and scores. Each of these will give you slightly different results, so it is worth taking your time searching.

The aim of these tools is to create a shortlist of colleges and courses that you can begin your more in-depth research on. While they may not exactly forecast your perfect college, they can eliminate some unsuitable ones.

Majors Offered

By now it is likely that you have a rough idea of what career or industry you would like to go into. For this, there is probably a specific major required. Using the Majors Offered filter on the college search tools is a great way to differentiate between an Engineering focused school or an Art-focused school, for example.

Location

Hate the cold? Totally okay! Your dream school isn't your dream school if it is in the middle of a snow zone. By filtering your college search by location, you can pinpoint some options on a certain coast or even in your home state.

Whether you prefer a rural, suburban, or city environment is important to consider, just as dorm or campus housing availability is.

Sports & Activities

While the study is important, college is about so much more than just your classes. Perhaps you are a star athlete or have a love for theatre? These are important factors in searching for a college, just like your SAT/ACT score. Remember, college is supposed to be a holistic experience, not just about what school looks good on paper.

Academic Credit

While your SAT/ACT scores may not be quite what you were hoping for to get into your dream school, there are sometimes other factors that affect getting into college. Perhaps you've taken AP classes, or have already earned some college credits to be transferred. It's important to take these into consideration when searching for a college- don't rule any school out too soon!

Price and Payment Support

It is no lie that college is expensive, but the wide range of costs can be confusing and prohibitive. Whether you have a certain budget in mind, are looking for an estimate to figure out a student loan, or are looking to research tuition support programs, it is important to factor these in right from the beginning of your search.

While no search engine can tell you your perfect college straight away, once you have your SAT/ACT scores, it may be easier to start narrowing down your choice of colleges. And remember: have fun! College is an exciting part of your life and research is just the first step.

COLLEGE FINANCING

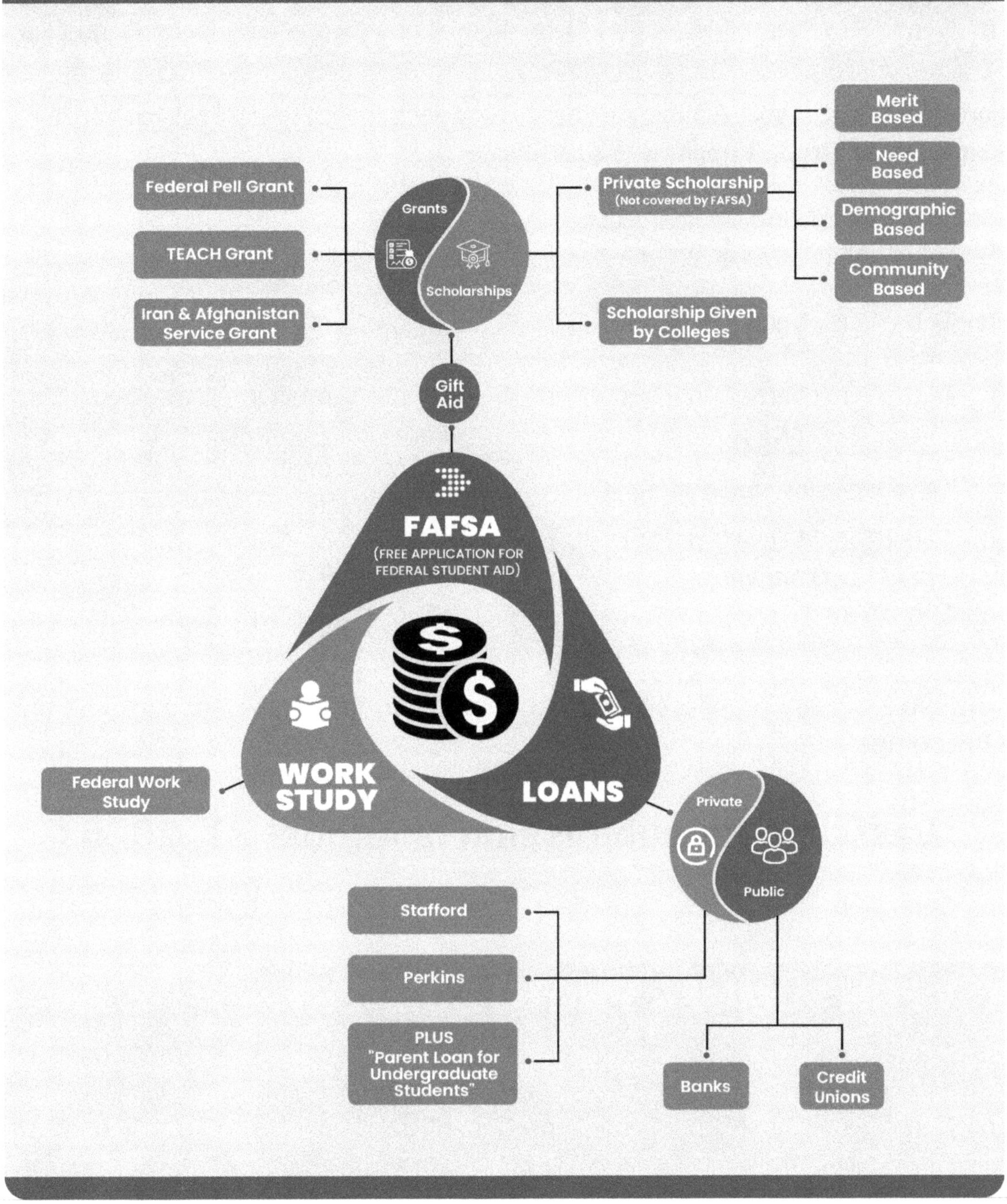

WEEK 10
LUMOS SHORT STORY & PHOTO CONTEST 2026

Write a short story based on your summer experiences and get a chance to win **$100 cash prize + 1 year free subscription to Lumos StepUp + trophy with a certificate**.

To enter the competition, follow the instructions.

STEP 1

Visit
www.lumoslearning.com/a/pld8 to register for online fun summer program.

STEP 2

After registering, log in to the student account and enter your summer story in Week 10 under "**Lumos Short Story Competition 2026**"

Note: If you have already registered, you can simply log in to your account and submit the story for the competition.

Last date for submission is August 31, 2026

WEEKLY FUN SUMMER PHOTO CONTEST

Take a picture of your summer fun activity and share it on Twitter or Instagram

Use the #SummerLearning mention

@LumosLearning on **Twitter**

@lumos.learning on **Instagram**

Tag friends and increase your chances of winning the contest.

PARTICIPATE AND STAND A CHANCE TO WIN $50 AMAZON GIFT CARD!

2025 Winning Story

Summertime in Sunny Asmara.

The summer before middle school, I traveled with my family to my home country of Eritrea. On June 20th, while my parents finished packing, I called my cousins Pitias and Abigail to tell them we'd be gone for two months. At first, they thought I was joking, but after convincing them, they wished me safe travels.I stuffed my bag with books, toys, and my tablet until it felt like I was carrying an elephant. Luckily, Uncle Natu helped us haul our luggage from the house to the airport. The airport process was long and boring, and after some Dunkin' Donuts, it was finally time to board. Flying with Turkish Airlines meant stopping in Istanbul first, and since I get motion sick on planes, the flights were unpleasant. At least the Burger King in Turkey was worth it, even if it cost us $80.

10 hours later (10 hours of eternity), and we finally landed in "Sweet Home, Asmara." I was exhausted, my leg was sore, and I had to share a bed with my mom, brother, and sister, but I was still excited for the adventures ahead.

In Eritrea, we did everything from playing soccer in the streets and in the stadium to riding horses and eating at fancy restaurants. We visited the salty beaches of Massawa, climbed the peaks of Senafe, and traveled 6 hours to my dad's hometown. Along the way, we met relatives and friends, saw baboons, mules, goats, donkeys, cows, and even hyenas at night. Climbing Senafe's mountains was so steep that I wondered how the heck we would even get down, but somehow we made it down on foot.

Even while having fun, I still made time for learning. Thanks to Lumos Learning, I was able to keep up with my schoolwork and felt well-prepared for middle school. This trip gave me amazing memories, time with family, and a stronger connection to where I come from. My name is Nftaliem, and that was my summertime in sunny Asmara.

Student Name: Nftaliem Tekle
Grade: 6

ANSWER KEY AND DETAILED EXPLANATION

WEEK 1

DAY 1

Question No.	Answer	Detailed Explanation
1	B	When combining polynomials, combine like terms by combining the coefficients. $(7x^3-5x^2-8x+15) + (-4x^3-2x^2+9x+6)$ $(7x^3+ -4x^3) + (-5x^2+ -2x^2) + (-8x+9x) + (15+6)$ $3x^3-7x^2+x+21$
2	B	When combining polynomials, combine like terms by combining the coefficients. Subtract $(3x^3 + x^2 - 9x)$ from $(8x^3 - 4x^2 - 3)$ means $(8x^3 - 4x^2 - 3) - (3x^3 + x^2 - 9x)$ $(8x^3 - 3x^3) + (-4x^2 - x^2) + (-(-9x)) + (-3)$ $5x^3 - 5x^2 + 9x - 3$
3	B	When multiplying polynomials, use modified distribution and the product rule for exponents. Then, combine like terms by combining the coefficients. $(x^2 - 6x + 5)(-4x^2 - x + 1)$ $x^2(-4x^2 - x + 1) - 6x(-4x^2 - x + 1) + 5(-4x^2 - x + 1)$ $-4x^4 - x^3 + x^2 + 24x^3 + 6x^2 - 6x - 20x^2 - 5x + 5$ $-4x^4 + (-x^3 + 24x^3) + (x^2 + 6x^2 - 20x^2) + (-6x - 5x) + 5$ $-4x^4 + 23x^3 - 13x^2 - 11x + 5$
4	D	Student must identify the like terms and combine by applying the rules of integer operations.
5	B	Student must identify the like terms and combine by applying the rules of integer operations; student must be careful to distribute the negative sign to the second polynomial.
6	B	The answer is B because its tone throughout is urging and demanding that we live meaningful lives.
7	A	The answer is A and can be found in the title of the poem
8	C	The correct answer is C because he is referring to people who are almost sleeplike. He is saying it puts our souls to sleep when we choose not to live life fully.

Question No.	Answer	Detailed Explanation
9	A	The correct answer is A. The following first line from stanza 7 makes it very clear. 'Lives of great men all remind us'.
10	A	The correct answer is A because the quote emphasizes this idea that our actions should make us better each day. That line suggests that we need to spend our days acting and doing so that tomorrow we are better people who have grown from that action.

WEEK 1

DAY 2

Question No.	Answer	Detailed Explanation
1	B	The Remainder Theorem is based on synthetic division, which is the process of dividing a polynomial f(x) by a polynomial D(x) and finding the quotient and remainder. This process evaluates the polynomial f(x) at a value where x = -c if D(x) = x + c. We can divide $(3x^3 - 3x^2 + 2x + 14)$ divided by (x-2) using synthetic division as: $\begin{array}{r\|rrrr} 2 & 3 & -3 & 2 & 14 \\ & & 6 & 6 & 16 \\ \hline & 3 & 3 & 8 & \boxed{30} \end{array}$ The remainder is the last number in the bottom row of the synthetic division. Therefore, f(2) = 30.
2	C	The Remainder Theorem is based on synthetic division, which is the process of dividing a polynomial f(x) by a polynomial D(x) and finding the quotient and remainder. This process evaluates the polynomial f(x) at a value where x = -c if D(x) = (x + c). We can divide$(x^3-6x^2+11x-6)$ by (x+2) using synthetic division as: $\begin{array}{r\|rrrr} -2 & 1 & -6 & 11 & -6 \\ & & -2 & 16 & -54 \\ \hline & 1 & -8 & 27 & \boxed{-60} \end{array}$ The remainder is the last number in the bottom row of the synthetic division. Therefore, the remainder is -60.
3	D	The Remainder Theorem is based on synthetic division, which is the process of dividing a polynomial f(x) by a polynomial D(x) and finding the quotient and remainder. This process evaluates the polynomial f(x) at a value where x = c if D(x) = x - c. We can divide$(x^3+2x^2+x-7)$by (x-2) using synthetic division as: $\begin{array}{r\|rrrr} 2 & 1 & 2 & 1 & -7 \\ & & 2 & 8 & 18 \\ \hline & 1 & 4 & 9 & \boxed{11} \end{array}$ The remainder is the last number in the bottom row of the synthetic division. Therefore, the remainder is 11.

Question No.	Answer	Detailed Explanation
4	C	The Remainder Theorem is based on synthetic division, which is the process of dividing a polynomial f(x) by a polynomial D(x) and finding the quotient and remainder. This process evaluates the polynomial f(x) at a value where x = c if D(x) = x - c. We can divide$(x^4+x^3-3x^2-4x-5)$by (x-2) using synthetic division as: 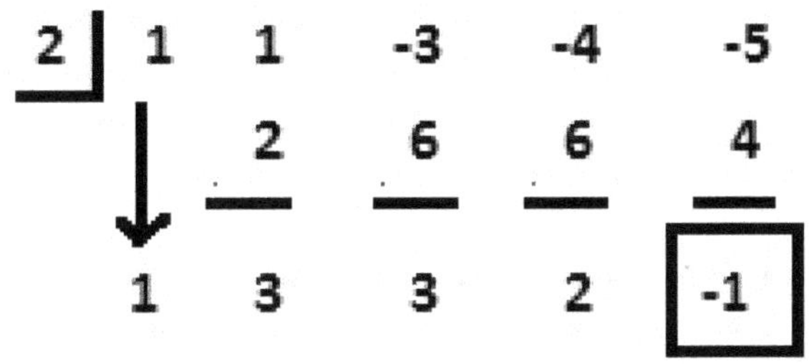 The remainder is the last number in the bottom row of the synthetic division and the function value. Therefore, h(2) = -1.
5	D	Student must substitute x = 1 and solve; if the substitution yields zero, there is no remainder; in this case $2(1)^3-5(1)+(1)-3=-5$.
6	D	Answer D- this is the best topic sentence as it provides what will be discussed and why.
7	C	Answer C- Although it is at the end of the paragraph, this is the topic sentence which states the main purpose of the writing.
8	B	Answer B- syntax refers to words arranged in phrases
9	B	The correct answer is B. When an informational text uses steps, that type of writing uses a sequence of events. The text structure is that of order and sequence.
10	D	Answer D- some of these can be used but the option D is the only one that shows definitive contrast.

Question No.	Answer	Detailed Explanation
1	B	The question asks us to find the zeros of $f(x) = (x^3+16x^2+64x)$. First, we will factor f(x) as x(x+8)(x+8) or $x(x+8)^2$. Next, we will set each factor equal to zero because the term "zeros of a function" means that the function has a value of zero: x+8=0, x=-8 and x=0. Thus, the function has two zeros, which are -8 and 0.
2	B	The question asks us to find the zeros of $h(x) = x^2 - 13x + 40$. First, we factor h(x) as (x - 5)(x - 8). Next, we will set each factor equal to zero because the term "zeros of a function" means that the function has a value of zero. x – 5 = 0, x = 5; x – 8 = 0, x = 8. Thus, the zeros of the function are 5 and 8.
3	B	Student must factor or use the quadratic formula to find the zeros of the function; student must also know that the solution to a quadratic equation is called a "root".
4	D	Student must set the equation equal to zero and then solve by factoring or using the quadratic formula to find the roots.
5	A	The student must identify that this quadratic has only one distinct root because it is a perfect square trinomial and therefore touches the x-axis only once.
6	B	The correct answer is B. Although the passage does inform, the use of description and details makes the main idea to describe the time.
7	A	Answer A is correct. Inquisition is a synonym of inquiry.
8	B	The answer is B. The author's purpose is strictly reflective at this point, citing background knowledge.
9	D	The correct answer is D. The essay draws on a variety of topics under the umbrella of baseball.
10	C	The correct answer is C. The writer is generalizing all boys which is a hasty generalization.

Question No.	Answer	Detailed Explanation
1	A	If we use the variable x for the number of cars Madison sells, and she sells two cars for every 10 bicycles, then she sells five times as many bicycles as cars. Thus, she sells 5x bicycles. If she sells four motor cycles for every car, the number of motorcycles she sells is 4x. The problem states that she makes 40 sales per month, so find the sum of the cars, bicycles, and motorcycles she sells and make that sum equal to 40. The equation is x+5x+4x=40.
2	D	The question states that Meghan ate two burgers and then gave each of her six friends three burgers. Let y be the number of burgers Meghan had. Then, based on the problem, the equation can be represented as y=2+6(3)=20. Therefore, Meghan had 20 burgers in the beginning.
3	A	Write an equation using the given information and prices with x representing the cost of a pack of paper. 7(0.85) + 5x + 8(1.20) + 15(0.45) = 25 Multiply and combine like terms. 22.3 + 5x = 25 Subtract 22.3 from both sides. 5x=2.70 Divide by 5:x = 0.54 Each pack of paper costs $0.54.
4	C	Each special contains four handkerchiefs, five pairs of socks, and two towels, which equates to 11 items. Create an equation that shows the product of the number of items in each special, the number of specials sold, and the average price of each item(x) equaling the total sales. 11 × 65x=2145. Multiply 11 × 65x = 715, so you have 715x = 2145. Divide both sides by 715 and the average price of the items in the special is $ 3.
5	C	Student must translate the written expression into an algebraic sentence and be careful to recall that the term "less than" reverses the order when written mathematically; subtraction is not commutative so order is important.
6	C	Answer C- Webster's dictionary has a citation to entry process for approval in the dictionary.
7	C	Answer C- the correct word should be "adverse".
8	A	Answer A- all the others have a direct effect on words changes over time.
9	D	Answer D- all the others are ways language can change over time.
10	C	Answer C- because 'by the birthday cake' describes the verb placed.

Question No.	Answer	Detailed Explanation
1	D	Begin by finding the slope of the line that contains the points. Use the points (6,9) and (8,29). The slope is $\frac{29-9}{8-6} = \frac{20}{2} = 10$. Next, use the point-slope formula of a line using the slope and the point (7,19). The formula is $y-y_1=m(x-x_1)$. With the slope and the point the equation is $y-19=10(x-7)$. Distribute and solve for y. $y-19=10x-70$; $y=10x-51$. Change the equation to function notation: $f(x)=10x-51$.
2	D	Begin by finding the slope of the line that contains the points. Use the points (6,19) and (8,11). The slope is $\frac{11-19}{8-6} = \frac{-8}{2} = -4$. Next, use the point-slope formula of a line using the slope and the point. The formula is $y-y_1= m(x-x_1)$. With the slope and the point the equation is $y-15=-4(x-7)$. Distribute and solve for y and $y-15=-4x+28$; $y=-4x+43$. Change the equation to function notation: $f(x)=-4x+43$
3	B	Begin by finding the slope of the line that contains the points. Use the points (24, 11) and (26, 21). The slope is $\frac{21-11}{26-24} = \frac{10}{2} = 5$. Next, use the point-slope formula of a line using the slope and the point (25, 16). The formula is $y-y_1=m(x-x_1)$. With the slope and the point the equation is $y-16=5(x-25)$. Distribute and solve for y and $y-16=5x-125$; $y=5x-109$. Change the equation to function notation: $f(x)=5x-109$.
4	B	Direct variation is represented by the equation $y = kx$, where k is the constant of variation; $k\neq0$; Therefore $20=k(4)$, and $k=5$.
5	B	The number of kilometers reached, k, is a function of 3 times the number of minutes, m. Therefore, $k=3m$.
6	A	Correct answer is A. Although answer B and C are good choices, A supports the overall theme of the passage.
7	C	Correct answer is C. The main idea of the passage is to learn by doing and experiencing nature.
8	A	Answer A, although incorrect, could be a tempting option. Wander is the best option in context of the sentence.
9	D	Answer D - varied sentence structure is part of the revision process.
10	D	Answer D - because 'while I finish making the dough' describes the verb start. It is a clause as opposed to a phrase because it contains both a subject and a predicate.

Question No.	Answer	Detailed Explanations
1	A	Create an inequality in which the cost at Hotel B is less than the cost at Hotel A. The cost at Hotel B is 747+16g where g is the number of guests. The cost at Hotel A is 431+38g. The inequality is 747+16g<431+38g. Subtract 431 from both sides and subtract 16g from both sides. This gives you 316<22g. Divide both sides by 22 and you get g>14.36. Therefore, if more than 14 people attend the party (that is, at least 15 people), Hotel B will be less expensive.
2	B	Create an inequality that shows the ticket revenue as greater than the expenses. This inequality is 8x≥420+2x where x is the number of tickets. The first step to solve the inequality is subtract 2x from both sides. This gives you 6x≥420. Divide both sides by 6 so that x≥70.The council must sell 70 tickets to recover the costs of the facility.
3	B	Create two equations, one that determines how many minutes Drake will need to type his paper and the other one that determines how many minutes Heidi will need to type her paper. Drake's equation is 45m=2160 and Heidi's equation is 60m=2640, where m represents the number of minutes. Solving each equation using division, you can see that Drake will take 48 minutes and Heidi will take 44 minutes to type their papers. Together it will take 92 minutes of typing to finish both term papers.
4	C	Create and solve an equation that sets Option A equal to Option B. Option A is 50m+200 where m is the number of months. Option B is 25m+500. The equation is 50m+200=25m+500. Solve the equation by subtracting 25m from both sides and subtracting 200 from both sides. The result is 25m=300. Divide both sides by 25 and m=12. If Zuhalie rents the bikes for 12 months, the cost of each bike will be the same.
5	B	Set up a proportion of guests to the total number of people, $\frac{8}{5} = \frac{x}{576}$. Solve by cross multiplying. 8x = 2880. Divide both sides by 8. So x=360.

Question No.	Answer	Detailed Explanations
6	A	The answer is A because no one is perfect and everyone will die.
7	D	The answer is D because he becomes completely focused and obsessed with the birthmark and cannot see anything beyond that.
8	D	The correct answer is D because he realizes the true significance of the birthmark.
9	D	The answer is D because theme, character development, and conflict are all a part of this quote.
10	A	The correct answer is A because it says “Many a gentleman would have risked his life for the honor of kissing that mysterious hand.”

Question No.	Answer	Detailed Explanations
1	B	Student must apply inverse operations by multiplying both sides of the equation by 2 and then dividing both sides by h to isolate b.
2	C	Student must apply the inverse operation of division to isolate h; dividing both sides of the equation by πr^2 will isolate h.
3	D	Student must apply the inverse operations: begin by multiplying both sides of the equation by R to isolate E^2 and then take the square root of both sides of the equation to isolate E.
4	A	Student must apply the inverse operations by first subtracting b^2 from both sides of the equation, which will isolate a^2. Next the student must take the square root of both sides of the equation to solve for a.
5	D	Student must know that the circumference formula of a circle is $C = 2\pi r$ and then solve it in terms of r.
6	C	Answer C- all other choices are specific to narrative writing.
7	C	The correct answer is C. First person is when the narrator uses I. For example, I was walking down the street when I saw it.
8	C	The correct answer is C. A dynamic character is one that grows and learns from the conflict by definition.
9	B	The correct answer is B, because quotes are used to offset dialogue from the narrative around it.
10	A	The correct answer is A, because first you set up the story with background information (exposition), then the action rises until the climax when it begins to fall and then you have the resolution or how the climax is fixed.

Question No.	Answer	Detailed Explanations
1	A	The report shows that in a high school with 400 students, there are 40% girls. The report says that 20% of the students in the school wear glasses. It is also known that 10% of the boys wear glasses. We are asked to determine how many boys wear glasses. If 40% of the students at the school are girls, then 60% are boys. This means there are 400 X 60% = 240 boys in the school. Then, since 10% of the boys wear glasses, 240 X 10% = 24 boys wear glasses.
2	C	The question says a study found that in a small town, 40% of all households have a pet and 35% have children living at home. The report also found that 60% of all households with children living at home have a pet. We are asked to find the probability that a household has no children living at home but has a pet if that household in the town is selected at random. "Change the whole answer to the suggestion as given and also change the answer choice in the section to B. - 40% of households have a pet, and 35% have children.60% of households with children have a pet, so 0.60×35%=21%.This means 21% have both children and a pet. So, 40%−21%=19%, which is closest to 19.5% (Choice B)." The answer is 26% of all families in the small town have no children living at home and a pet.
3	B	To find W given X we put the number of data points that are in W and also in X on top, then we put the total number of data points that are in X on bottom. This gives us $\frac{2}{12}$ which reduces to $\frac{1}{6}$.
4	C	To find the probability of Y or Z given X, we add up the data points that are in Y or Z and also in X, that would be 5 and it goes on the top. On bottom, put the total number of data points in X. $\frac{5}{12}$
5	A	To find the probability of X or Y given Z, we add up the data points that are in X or Y and also in Z, that would be 4 and it goes on the top. On bottom, put the total number of data points in X, then reduce. $\frac{1}{3}$

Question No.	Answer	Detailed Explanations
6	B	Answer B- Although he admits not knowing for sure, the author provides a cause and effect that allows the reader to draw his or her own conclusions.
7	B	Answer B- The author is clear that the slight changes in each species make for endless possibilities of change.
8	B	The correct answer is B because the reader has to know the thesis in order to be able to identify the ways the writer presents the work.
9	D	The correct choice is D because the others are simply level 1 analysis that don't require implicit reading.
10	D	The correct answer choice is D because the writer has to be sure that the evidence is connected to the point and cite his/her sources

Question No.	Answer	Detailed Explanations
1	C	The question states that a is the independent variable. Therefore, the function must be in terms of a giving you the function notation of f(a). Notice in the table that b is always 4 times a, so the function is f(a)=4a.
2	A	The questions states that the relationship is expressed as ordered pairs (c,d). This means that d is the output of a function of c. The question states that the output is always 9 more than twice the input variable. The input variable is c, twice the input variable is 2c, and 9 more than means add 9. Therefore, the function is f(c)=2c+9.
3	A	The questions states that the relationship is expressed as ordered pairs (m,n). This means that n is the output of a function of m. The question states that the output is always 2 less than three times the cube of the input variable. The input variable is m, the cube of m is m^3, three times the cube of m is m^3, and 2 less than means subtract 2. Therefore, the function is $f(m)=3m^3-2$.
4	C	The formula for exponential decay is $y=ab^x$ where a is the initial value and b is the growth decay factor. When b>1, the function is an exponential growth. When $0 < b < 1$, the function is an exponential decay function.
5	C	When solving for the final velocity of an object with a known initial velocity and a constant acceleration over time, we use the formula final velocity = initial velocity + (acceleration × time). So answer choice C is correct.
6	C	The correct answers are B and C. We should use a hyphen with prefixes or suffixes like anti. Here 'anti' is a prefix in the word anti-inflammatory. We should also hyphenate when expressing ages as adjectives before a noun. Here, in the option C, the six-year-old appears before the noun "girl".
7	A	Answer A- notify is the verb form
8	B	Answer B- because it is a contraction meaning 'they are.'
9	B	Answer B because it separates two independent clauses.
10	C	Answer C because it separates the dependent clause Soon after moving to a new house from the independent clause Thomas and Denny made friends with their neighbors.

Question No.	Answer	Detailed Explanations
1	B	Arc length = rθ; therefore length = $(12ft)((\frac{7}{4})\pi)=21\pi$ ft
2	A	Arc length = rθ ; therefore length =$(16ft)((\frac{\pi}{6}))=\frac{8\pi}{3}$ ft.
3	C	Arc length = rθ ; therefore length = $(18ft)((\frac{2}{3})\pi)=12\pi$ft.
4	A	Area =$(\frac{\theta}{2\pi})\pi r^2= (\frac{\theta}{2})r^2$; therefore Area = $(\frac{\frac{\pi}{6}}{2})(10\ cm)^2=\frac{25\pi}{3}$ πcm².
5	C	Area = $(\frac{\theta}{2\pi}).\pi r^2=\frac{\theta}{2}r^2$; therefore Area=$(\frac{\frac{2\pi}{3}}{2})(18\ cm)^2=108\pi cm^2$.
6	C	The text structure here is to describe. Hence, the correct answer is C.
7	B	The correct answer is B. The passage gives detailed information about anesthesia and its effect on different parts of the brain.
8	B	The correct answer is B- capitalizing key words was a technique the author used to highlight key words and summon emotion from the reader.
9	A	The correct answer is A. When you are writing to inform or explain, you should include many strong facts. Writing to inform and explain is about choosing important facts and then developing your writing around that. Opinion and writing in first person has little place in an informative piece.
10	D	Answer D, because a relative clause is not a complete sentence and it gives extra information about the preceding noun.

Question No.	Answer	Detailed Explanations
1	B	The question asks you to solve the radical equation $\sqrt{x}=0.9$. Solve this equation by squaring both sides.$(\sqrt{x})^2=(0.9)^2$. Squaring cancels the square root, so the answer is $x=0.81$.
2	D	The question asks you to solve the rational equation $\frac{2}{x+6}=\frac{6}{x-4}$. This equation is a proportion and can be solved by cross multiplication. This gives you $2x-8=-6x-36$. Next, add 6x to both sides and add 8 to both sides, resulting in $8x=-28$. Now, divide both sides by 8 and $x=\frac{-28}{8}=\frac{-7}{2}$.
3	C	The question asks you to solve the radical equation $\sqrt{x}=0.36$. Solve this equation by squaring both sides. $(\sqrt{x})^2=(0.36)^2$. Squaring cancels the square root, so the answer is $x=0.1296$.
4	D	The question asks you to solve the rational equation $\frac{9}{z+3}=\frac{-3}{z-6}$. This equation is a proportion, so solve it by cross multiplying. This gives you $9z-54=-3z-9$. Next, add 3z to both sides and add 54 to both sides, resulting in $12z=45$. Now, divide both sides by 12 and $z=\frac{45}{12}=\frac{15}{4}$.
5	A	Student must square both sides of the equation (apply the inverse operation) and then solve the one-step equation for a.
6	C	The answer is C. The author describes that people tend to accept what has been handed down to them instead of further exploring the source of their convictions or beliefs.
7	C	The answer is C. Throughout the text; the author refers to the tendency of people to blindly accept information as true multiple times, which is why he encourages more thinking about our thinking.
8	D	The answer is D. The adapted characters are the pigs and the wolf. The theme remains the same (protecting oneself), and the antagonist is still the wolf as evident by the words huff and puff.
9	E	The answer is E because these are not elements that you would see easily if you set it in New York City in 2011. They would be replaced with other items that are more appropriate to a more modern setting.
10	D	The answer is D. All of those might be clues that the story has been adapted from another source.

WEEK 3

DAY 2

Question No.	Answer	Detailed Explanations
1	A	The question states that the furniture store chain is a national company. That means its customers live in all fifty states. The survey was conducted using only adults from Seattle, Washington. The sample was selected from only a very small portion of the country, and therefore does not represent all potential customers, thus, it will not produce a valid conclusion.
2	A	Voluntary sampling involves people self-selecting themselves to participate in the survey. Thus, people passing the display can decide whether or not they want to be surveyed. Often, only people with a strong interest in the main topic being surveyed participate in the survey.
3	B	The population of the survey is the group of people that the survey participants represent. For example, the 1397 students in the school are the population. Every member of the population should have an equal opportunity to participate in the survey.
4	B	Obtaining data using surveys has the following advantages: (1) Surveys allow researchers to collect a large amount of data in a relatively short period of time. (2) Surveys are less expensive than many other data collection techniques. (3) Surveys can be created quickly and administered easily. (4) Surveys can be used to collect information on a broad range of things, including personal facts, attitudes, past behaviors, and opinions.
5	A	The question states that the grocery store chain is a national company. That means its customers live in all fifty states. The survey was conducted using only adults from Chicago, Illinois. The sample was selected from only a very small portion of the country, and therefore does not represent all potential customers; thus, it will not produce a valid conclusion.
6	B	Answer B. C and D could help but are not the most effective. A is the opposite of the correct answer.
7	A	The answer is A, because you would tell a story using metaphors and sensory details (details that try to get the reader to see, smell, hear, taste, and touch).

Question No.	Answer	Detailed Explanations
8	D	The correct answer is D because these are all types of informational texts. Sequence is a list of information and B and C deal with informing the reader of problems /causes and solutions/effects.
9	B	The correct answer is B, because the last sentence, which of the thesis, says: "While there is an argument that 16 and 17-year-olds are too immature to vote, there is also a danger that they might not vote at all.
10	B	The answer is B. You can tell this because the writer hints that, since the voting age is 18, teenagers may never become voters, and that is the problem.

<table>
<tr><th>Question No.</th><th>Answer</th><th>Detailed Explanations</th></tr>
<tr><td>1</td><td>C</td><td>The question asks us to provide the range of $g(x)=x^2-2x+5$ on the domain {1,2,3,4}. The domain of a function is the set of input values into the function. The range of the function is the set of output values for the domain. This table shows the domain of g(x) and the calculations of the range for each element of the domain.

<table>
<tr><th>Domain</th><th>Calculations</th><th>Range</th></tr>
<tr><td>1</td><td>$(1)^2-2(1)+5$</td><td>4</td></tr>
<tr><td>2</td><td>$(2)^2-2(2)+5$</td><td>5</td></tr>
<tr><td>3</td><td>$(3)^2-2(3)+5$</td><td>8</td></tr>
<tr><td>4</td><td>$(4)^2-2(4)+5$</td><td>13</td></tr>
</table></td></tr>
<tr><td>2</td><td>A</td><td>The question asks us to provide the range of

$f(x)=\frac{(x^2-x-6)}{x-3}$

on the domain {10,11,12,13,14}. The domain of a function is the set of input values into the function. The range of the function is the set of output values for the domain. This table shows the domain of f(x) and the calculations of the range for each element of the domain.

<table>
<tr><th>Domain</th><th>Calculations</th><th>Range</th></tr>
<tr><td>10</td><td>$\frac{(10^2-10-6)}{10-3}$</td><td>12</td></tr>
<tr><td>11</td><td>$\frac{(11^2-11-6)}{11-3}$</td><td>13</td></tr>
<tr><td>12</td><td>$\frac{(12^2-12-6)}{12-3}$</td><td>14</td></tr>
<tr><td>13</td><td>$\frac{(13^2-13-6)}{13-3}$</td><td>15</td></tr>
<tr><td>14</td><td>$\frac{(14^2-14-6)}{14-3}$</td><td>16</td></tr>
</table></td></tr>
</table>

<table>
<tr><th>Question No.</th><th>Answer</th><th>Detailed Explanations</th></tr>
<tr><td>3</td><td>B</td><td>The question asks us to provide the range of $g(x)=3x-23$ on the domain {15,18,25,27}. The domain of a function is the set of input values into the function. The range of the function is the set of output values for the domain. This table shows the domain of $g(x)$ and the calculations of the range for each element of the domain.
<table>
<tr><th>Domain</th><th>Calculations</th><th>Range</th></tr>
<tr><td>15</td><td>3(15)-23</td><td>22</td></tr>
<tr><td>18</td><td>3(18)-23</td><td>31</td></tr>
<tr><td>25</td><td>3(25)-23</td><td>52</td></tr>
<tr><td>27</td><td>3(27)-23</td><td>58</td></tr>
</table></td></tr>
<tr><td>4</td><td>C</td><td>The question asks us to provide the range of $h(x)=4x^2-23x+8$ on the domain {1,3,5,7,9}. The domain of a function is the set of input values into the function. The range of the function is the set of output values for the domain. This table shows the domain of $h(x)$ and the calculations of the range for each element of the domain.
<table>
<tr><th>Domain</th><th>Calculations</th><th>Range</th></tr>
<tr><td>1</td><td>$4(1)^2-23(1)+8$</td><td>-11</td></tr>
<tr><td>3</td><td>$4(3)^2-23(3)+8$</td><td>-25</td></tr>
<tr><td>5</td><td>$4(5)^2-23(5)+8$</td><td>-7</td></tr>
<tr><td>7</td><td>$4(7)^2-23(7)+8$</td><td>43</td></tr>
<tr><td>9</td><td>$4(9)^2-23(9)+8$</td><td>125</td></tr>
</table></td></tr>
<tr><td>5</td><td>A</td><td>The question asks us to provide the range of $f(x)=\frac{(2x^2-13x-7)}{(2x+1)}$ on the domain {2,6,10,14}. The domain of a function is the set of input values into the function. The range of the function is the set of output values for the domain. This table shows the domain of $f(x)$ and the calculations of the range for each element of the domain.
<table>
<tr><th>Domain</th><th>Calculations</th><th>Range</th></tr>
<tr><td>2</td><td>$\frac{2(2)^2-13(2)-7}{2(2)+1}$</td><td>-5</td></tr>
<tr><td>6</td><td>$\frac{2(6)^2-13(6)-7}{2(6)+1}$</td><td>-1</td></tr>
<tr><td>10</td><td>$\frac{2(10)^2-13(10)-7}{2(10)+1}$</td><td>3</td></tr>
<tr><td>14</td><td>$\frac{2(14)^2-13(14)-7}{2(14)+1}$</td><td>7</td></tr>
</table></td></tr>
</table>

Question No.	Answer	Detailed Explanations
6	A	One can assume it means large and powerful because the enemy appears so great, as shown by the urgency of his tone and the lengths he is going to in order to persuade the people to fight.
7	A	The answer is A because sarcasm is the use of irony to emphasize disgust for a subject.
8	B	He wants his people to rise up and revolt against British rule.
9	B	He uses B, metaphor, to compare life under British rule to slavery. He says by not fighting we purchase peace, but the price is chains. This is a comparison between two ideas.
10	A	The correct answer is A because all three are used. Hyperbole is exaggerated language, and the description of the chains is exaggerated. The metaphor is the chains as compared to their "enslavement" to the British.

Question No.	Answer	Detailed Explanations
1	B	A line is defined as a set of points on a plane that extend in two directions without ending. Thus, a line is one-dimensional. It has a length, but zero width. If you draw a line with a pencil, and examine it with a microscope, it would show that the pencil mark has a measurable width. The pencil line is just a way to illustrate the idea on paper. In geometry however, a line has no width.
2	B	A plane is a set of points on a flat surface or a 2-dimensional object, stretching to infinity in all directions. In geometry, a plane is a flat two-dimensional "surface", but with no thickness and no finite length or width. To define a plane, we must have three points, each of which forms a line with the other two points within the plane. When we illustrate a plane, we make a shape like a sheet of paper, but the plane has an infinite length and width.
3	A	In geometry, a circle is the set of all points (on the plane) that are an equal distance from a given point called the center. A circle is named by its center. Thus, the circle with its center at point A is called circle A. The words "on the plane" are necessary for the correct definition. Without these words, we are describing a sphere.
4	D	Two lines that form a T are perpendicular to each other. The two lines intersect at a right angle and their slopes are opposite reciprocals. Perpendicular is used to describe lines, angles and direction. In geometry, a right angle is 90 °, forming a perfect L. We indicate that two lines are perpendicular with a small box at their intersection point. On a plane, horizontal lines and vertical lines are perpendicular to each other.
5	C	In geometry, a solid object is a set of points that forms a figure that has three dimensions. The three dimensions are called width, depth and height. Some examples of solids include cubes, pyramids, spheres, cylinders, and prisms.

Question No.	Answer	Detailed Explanations
6	A	Answer A- The correct spelling is 'definitely.'
7	C	Answer C: this is the only adjective form of compete.
8	A	Answer A because it introduces a list of items.
9	D	Answer D because it means 'extremely,' which is the best fit for the context of the sentence; to indicates direction; two is a number; and toe is a part of the body.
10	A	Answer A because a colon is used between two independent clauses only when the second further explains or elaborates on the first.

Question No.	Answer	Detailed Explanations
1	C	If we rearrange the data set in the question in numerical order, we have this set. {3,3,3,4,4,5,5,5,5,6,6,6,7,7,8,8,8}. Therefore, the line plot must contain three 3's,two 4's, four 5's, three 6's, two 7's, and three 8's. The dot plot below contains the correct data set. 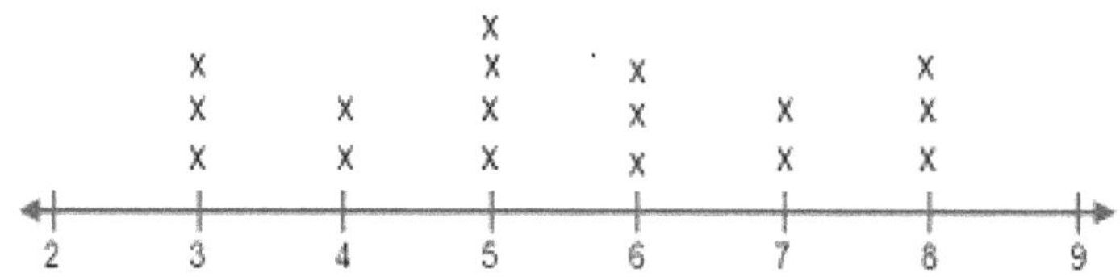
2	C	A box and whiskers plot provides five critical numbers in a data set. These numbers are marked with a square in the image below. 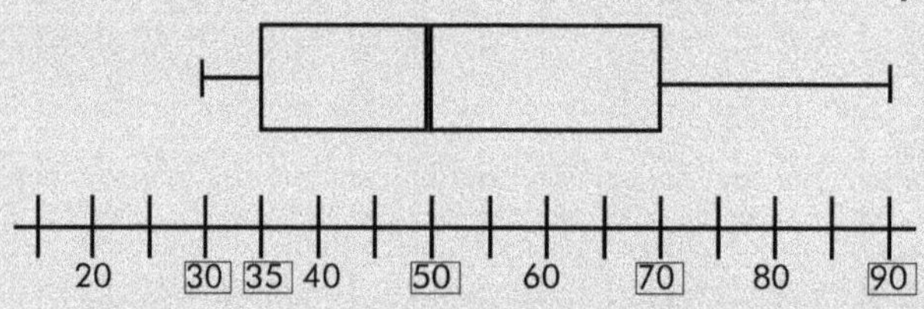 From left to right, these five numbers are: (i) the smallest number in the data set: 30; (ii) the first quartile, which is the median of the lower half of the data set: 35; (iii) the median of the data set: 50 ; (iv) the third quartile, which is the median of the upper half of the data set: 70; and (v) the largest number in the data set: 90.
3	C	In the histogram, the x-axis gives the values in the data set. The y-axis gives the quantity of each value in the data set. Based on the histogram, the data set contains four 2's, six 4's, four 6's, and two 8's.
4	A	A box and whiskers plot provides five critical numbers in a data set. These numbers are marked with a square in the image below. 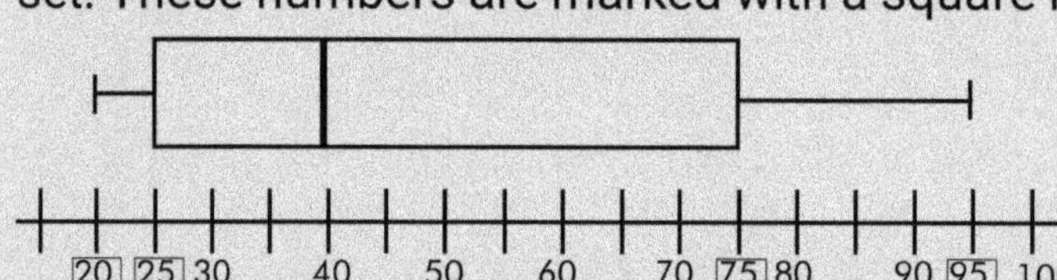 From left to right, these five numbers are: (i) the smallest number in the data set: 20; (ii) the first quartile, which is the median of the lower half of the data set: 25 ; (iii) the median of the data set: 40 ; (iv) the third quartile, which is the median of the upper half of the data set: 75 ; and (v) the largest number in the data set: 95.

Question No.	Answer	Detailed Explanations
5	B	Consider each option. Option A -- There are no data points in the 60 column, this statement is false. Option B -- The 90 column has more data points than any other column, therefore most students scored 90. This option is correct. Option C-- Since no students scored 60, this would be the score with the fewest student scores. Option C is incorrect. Option D -- There were 2 students scoring 70 and 5 students scoring 80 for a total of 7 students. There are 7 data points in the 90 column as well, meaning there were equal number of students scoring either 70 or 80 and students scoring 90. Option D is not correct.
6	A	This statement is the strongest support for the main idea and clearly shows that the direction of traveling enhances the view.
7	B	Answer B- the main idea is expressed in the thesis statement. They are not the same, as there can be other statements throughout the piece that also support the main idea.
8	B	The correct answer is B. The inciting incident is what causes the action to rise.
9	B	The correct answer is B because facts are central to an informative piece of writing.
10	B	Answer B, because a comma separates the dependent clause, starting with "Rosa Parks' refusal to sit in the back of the bus," from the independent clause, "the Montgomery bus boycott brought national attention to the issue of segregation."

WEEK 4

DAY 1

Question No.	Answer	Detailed Explanations
1	D	The question asks you to find the solution to the equation $\frac{x}{5}$- 11=2.Begin by adding 11 to both sides of the equation. This gives you $\frac{x}{5}$=13. Next, multiply both sides by 5 and x=65.
2	B	The question asks you for the solution to the inequality $5>\frac{(x+16)}{7}$. Begin by multiplying both sides of the inequality by 7. This changes the inequality to $35>x+16$. Next, subtract 16 from both sides, leaving $19>x$. Since the variable is normally on the left side of an inequality, change the inequality to $x<19$.
3	C	The question asks you to find the solution to the equation 10x-5x-4=9. Begin by combining like terms on the left side of the equation. This gives you 5x-4=9. Next, add 4 to both sides, changing the equation to 5x=13. Last, divide both sides by 5 and x=$\frac{13}{5}$.
4	B	The question asks you for the solution to the inequality $6<\frac{(y-12)}{6}$. Begin by multiplying both sides of the inequality by 6. This changes the inequality to $36<y-12$. Next, add 12 to both sides, leaving $48<y$. Since the variable is normally on the left side of an inequality, change the inequality to $y>48$.
5	D	The question asks you for the solution to the inequality $\frac{3\geq(x+7)}{8}$. Begin by multiplying both sides of the inequality by 8. This changes the inequality to $24\geq x+7$. Next, subtract 7 from both sides, leaving $17\geq x$. Since the variable is normally on the left side of an inequality, change the inequality to $x\leq17$.
6	A	The correct answer is A. He constantly remarks on their state of “slumber”, as though they are sleeping through this oppression.
7	D	The correct answer is D. He makes reference to fear, slumber, and the anarchy of destiny in the poem.
8	A	The correct answer is A. He feels that they are sleeping and need a wake-up call.
9	A	The correct answer is A. The master of show is whoever is leading the people.
10	A	The answer is a because the poet uses “I” to convey his ideas

Question No.	Answer	Detailed Explanations
1	B	The question asks us to provide the difference between the means of two sets of data. We find a mean by adding the numbers and then dividing by the number of numbers in the set. First, we will find the mean of set A: $\frac{(2+4+6+8+10+12)}{6}=\frac{42}{6}=7$. Next, we will find the mean of set B: $\frac{(3+5+7+9+11+13)}{6}=\frac{48}{6}=8$. The difference between the two means is 1.
2	B	The question asks us to find the difference in the medians of Set A and Set B. The median of a set of numbers is the center number when the numbers are in order from least to greatest. If there is an even number of numbers in the set, then the median is the average of the two middle numbers in the set. Both sets are displayed as dot plots, which show the quantity and value of each member of the set. The dot plot places them in order from least to greatest. Therefore, we must identify how many numbers are in each set. Both sets have 30 numbers, so we find the 15th and 16th numbers in each set, and use their average as the median. In Set A, the 15th and 16th numbers are both 5. Thus, their average is 5. In Set B, the 15th and 16th numbers are both -5. Thus, their average is -5. thus, their average is -5. The difference between these medians is 10.
3	C	The question asks us to provide the difference between the medians of two sets of data. We find a median by arranging the numbers in the set from the least to the greatest. The median is the middle number. First, we will find the median of set A: {6,8,10,12,14,16,18,20,22}. The numbers are already in order from least to greatest. Notice that the set has 9 numbers, so the middle number is the fifth number, which is 14. Next, we will find the median of set B: {3,5,7,9,11,13,15,17,19}. The numbers are, again, already in order from least to greatest. Notice that the set also has 9 numbers, so the middle number is also the fifth number, which is 11. The difference between the two medians is 3.

Question No.	Answer	Detailed Explanations
4	D	A box and whiskers plot provides five critical numbers in a data set. From left to right, these five numbers are: (i) the smallest number in the data set; (ii) the first quartile, which is the median of the lower half of the data set ; (iii) the median of the data set ; (iv) the third quartile, which is the median of the upper half of the data set ; and (v) the largest number in the data set. The median of Set A is 35 and the median of Set B is 60. The difference between these two medians is 25.
5	C	A geometry theorem states that if a line is tangent to a circle at the outer endpoint of a radius, then the radius is perpendicular to the tangent line. Thus, ΔOQR is a right triangle and the relationship between the sides and the hypotenuse must obey the Pythagorean Theorem. This means that $(OQ)^2 + (QR)^2 = (OR)^2$, so $x^2 + 15^2 = (x + 9)^2$. Solve the equation $x^2 + 15^2 = (x + 9)^2$. Using correct order of operations, perform the exponents in the equation, giving $x^2 + 225 = x^2 + 18x + 81$. Cancel the x^2 on each side, leaving $225 = 18x + 81$. Subtract 81 from both sides and divide both sides by 18. The solution to the equation is $x=8$.
6	D	The correct answer is D, because the writer can do all of those things
7	D	The answer is D, because all changes are needed.
8	C	The correct answer is C and the dependent clause is "Because I love dogs."
9	B	This statement is false. The writer should not be drafting using informal language. That can appear in the pre-write, but by the time the writing is in drafting, the language should be more formal.
10	B	The answer is B. Your writing has to have different types of sentences in it so the writing is varied and the reader stays interested.

Question No.	Answer	Detailed Explanations
1	B	The median of a data set is the middle number of the data set when the numbers are arranged from least to greatest. If the data set has an even number of members, then the median is the average of the middle two numbers. Here are the data sets arranged in order from least to greatest. Class A: 40, 50, 55, 60, 65, 68, 70, 77, 78, 80, 84, 88, 95 Class B: 50, 58, 59, 60, 67, 68, 74, 78, 80, 81, 91, 92, 95 Each class has 13 students, so the seventh position in the set, when in order, is the median. The median of Class A is 70 and the median of Class B is 74. The difference between the medians is 4.
2	D	Perform a statistical analysis of the two groups. For Stacey's phone calls mean length is μ=34.33 minutes and median is x=32.5 minutes. For Dianna's phone calls mean length is μ=35.66 minutes and the median is x=33 minutes. The set of data clearly shows that Stacey made a call that was 13 minutes long. When we compare the median call length for each person, Stacey's median call length is 32.5 minutes and Dianna's median call length is 33 minutes, so Dianna's median call length is longer. When we compare the mean and median in each data set, we can see that the median is smaller than the mean in both sets of data. The correct answer is II and III.
3	D	When we have data displayed in a stem plot, the numbers in the data set are already in order from least to greatest. The median of a data set is the middle number in the set. One way to find the middle number is to eliminate the smallest number and the largest number. Then eliminate the next smallest number and the next largest number. Repeat until there is only one or two numbers left in the middle. The figure below shows the result in this question.

Mr. Smith		Ms. Jones
~~8~~	0	~~1~~
~~1~~ ~~2~~ ~~3~~	1	~~3~~
8	2	~~8~~
2 ~~3~~ ~~4~~	3	~~3~~ 4
~~9~~	4	5 ~~6~~ ~~7~~
~~7~~	5	~~2~~ ~~3~~

Question No.	Answer	Detailed Explanations
3 cont.		If there is only one number left, that number is the median. If there are two numbers left, the median is the average of those two numbers. In Mr. Smith's class, 28 and 32 are left, so the median is $\frac{(28+32)}{2}=30$ In Ms. Jones' class, 34 and 45 are left, so the median is $\frac{(34+45)}{2}=39.5$ The sum of these medians is 69.5.
4	B	Perform a statistical analysis of the two groups. For Survey I, the mean number of water bottles is μ = 36.7 bottles per month. For Survey II, the mean number of water bottles is μ = 44.2 bottles per month. The average of the means is $\frac{(36.7+44.2)}{2}=40.45$
5	A	Perform a statistical analysis of the two groups. For the residents in Hifalutin Garden, the mean age is μ=86.2 years. For the residents in Pretentious Terrace, the mean age is μ=70.2 years. Since the means are significantly different, we can say that the average age that a patient shows early signs of dementia at Hifalutin Garden are higher than at Pretentious Terrace. The mean age at which patients show early signs of dementia at Hifalutin Garden is much higher than the average age in Pretentious Terrace. Therefore, this information supports the idea that regular mental stimulation activities do delay the age at which people show signs of dementia.
6	A	The correct answer is A because it shows the "phenomenon" and the writer presents it as a good thing for our culture.
7	A	The answer is A because the writer presents a balanced argument.
8	D	The correct answer is D - the text structure shows the effects of what will happen by marrying specific types.
9	A	The Correct answer is A - this is written in a narrative text structure.
10	C	The Correct answer is C - incorporating questions. This was done to keep the thoughts going of the reader and to have them thinking about the effects he is discussing.

Question No.	Answer	Detailed Explanations
1	B	The input variable is x. Cube rooting the input variable is a term $\sqrt[3]{x}$. Doubling that value is a term $2\sqrt[3]{x}$. Adding 9 is the term +9. Put together, we have $f(x)=2\sqrt[3]{x}+9$. Therefore, $f(343)=2\sqrt[3]{343}+9=2(7)+9=14+9=23$
2	D	The input variable is x. Multiplying the input by 5 gives us 5x. Doubling that answer gives us the term 2(5x). Adding 1 gives us the term +1. Put together, we have $h(x)=2(5x)+1$. Therefore, $h(12)=2[5(12)]+1=120+1=121$.
3	D	The input variable is x. Twice the input is 2x. Cubing that value is the term $(2x)^3$. Adding three times the input is a term +3x. Then, subtracting 1 is the term -1. Put together, we have $f(x)=(2x)^3+3x-1$. Therefore, $f(3)=2(3)^3+3(3)-1=216+9-1=224$
4	B	Find the function value at x = -4 by plugging in a -4 for x as shown below. Then our point will have an x-value of -4 and we determine a y-value by substitution. $f(x) = -3x + 5$ $f(-4) = -3(-4) + 5$ $f(-4) = 12 + 5 = 17$ This is the point (-4, 17), which is answer choice B.
5	D	Find the function value at x = -6 by plugging in a -6 for x as shown below. Then our function notation will start with f(-6) because we are evaluating the function at -6 and it will be equal to the value we get when we input -6. $f(x)=(\frac{2}{3})x-4$ $f(-6)=(\frac{2}{3})(-6)-4$ $f(-6)=-4-4=-8$ So the answer in function notation is $f(-6)=-8$. This shows an input of -6 and an output of -8.
6	A	The answer is A. This sentence is simple, containing only one independent clause. The rest of the sentences have an independent and a dependent clause joined by a subordinating conjunction.
7	D	The answer is D. Both ways of spelling the word are considered correct. Grey is a variant spelling of gray.
8	D	Syntax refers to the order of words in a sentence. It is the way in which words are put together to form phrases, clauses, or sentences.

Question No.	Answer	Detailed Explanations
9	B	The answer is B. This sentence has the required independent and dependent clauses joined by a subordinating conjunction.
10	D	Answer D because The New York Times, the title of a newspaper, is listed in the source title position in the Works Cited entry.

Question No.	Answer	Detailed Explanations
1	D	The space left in the can is the difference between the volume of the can and the volume of the sphere according to: $\Delta V=\pi r^2h-(\frac{4}{3})\pi r^3$ $\Delta V=\pi(6cm)^2(10cm)-(\frac{4}{3})\pi(6cm)^3$ $\Delta V=72\pi cm^3$
2	C	The fraction of the volume occupied by the cone is the ratio of the volume of the cone to the volume of the can. $(\frac{\frac{1}{3}\pi r^2h}{\pi r^2h})=\frac{1}{3}$
3	C	The total volume of the object is the sum of the volumes of the hemisphere and the cone. $V = ½ (\frac{4}{3}) \pi r^3+(\frac{1}{3})\pi r^2h$ $V = ½ (\frac{4}{3}) \pi(3cm)^3+(\frac{1}{3})\pi(3cm)^2(10cm)V=48\pi cm^3$
4	C	Use the formula for the volume of a pyramid: V = ½ × a × c × h. In this case the length is 15 cm, the base is 10 cm, and the height is 9 cm. Therefore: V = ½ × 15 × 10 × 9 = 675 cm^3.
5	A	The formula for the volume of a cone is $V=\pi r^2(\frac{h}{3})$. The height (h) is given as 2 feet. The radius (r) can be found by dividing the diameter in half. The radius is then $\frac{4m}{2}$=2m. These values can then be substituted into the formula. $V=\pi r^2(\frac{h}{3})$ $V=(3.14)(2m)^2(\frac{2m}{3})=(3.14)(4m^2)((\frac{2}{3})m)=8.373m^3$.
6	C	The text structure here is to describe. Hence, the correct answer is C.
7	A	Correct answer is A. Although it could be reference, the tone of the author's writing is from a more personal standpoint.
8	D	The answer is D. D does not have any sensory detail in its answer. The remaining answers relate to one of the senses.
9	D	The correct answer is D because they all will inform the reader of something
10	A	The correct answer is A. The underlined part of is a complete sentence with its own subject and predicate.

Question No.	Answer	Detailed Explanations
1	B	The question asks you to find the solutions to the quadratic equation $2x^2-3x+1=0$. Find the solutions using factoring. In factoring, you factor the first and the third term in such a way that the two factors add to give the middle term. Since the third term is positive, the two factors of that term must have the same sign. Since the second term is negative both of the factors must be negative. Thus, the factored equation is $(2x-1)(x-1)=0$. Find the solutions by making each factor equal to zero. $2x-1=0$; $x= \frac{1}{2}$ and $x-1=0$; $x=1$.
2	D	The question asks you to find the solutions to the quadratic equation $2x^2+8x+6=0$. Find the solutions using factoring. In factoring, you factor the first and the third term in such a way that the two factors add to give the middle term. Since all of the terms can be divided by 2, simplify the quadratic equation to $x^2 + 4x + 3 = 0$. Also, since all of the terms are positive, the two factors of that term must have a positive sign. Thus, the factored equation is $(x+1)(x+3)=0$. Find the solutions by making each factor equal to zero. $x+1=0$; $x=-1$ and $x+3=0$; $x=-3$.
3	A	The question asks you to find the solutions to the quadratic equation $4x^2-11x+6=0$. Find the solutions using factoring. In factoring, you factor the first and the third term in such a way that the two factors add to give the middle term. Since the third term is positive, the two factors of that term must have the same sign. Since the second term is negative both of the factors must be negative. Lastly, since the middle term is odd, the factors of the first term must be odd and even. Thus, the factored equation is $(4x-3)(x-2)=0$. Find the solutions by making each factor equal to zero. $4x-3=0$; $x=\frac{3}{4}$ and $x-2=0$; $x=2$.
4	B	The question asks you to find the solutions to the quadratic equation $2x^2-3x+1=0$. Find the solutions using factoring. In factoring, you factor the first and the third term in such a way that the two factors add to give the middle term when checking the factoring using the FOIL process. Since the third term is positive, the two factors of that term must have the same sign. Since the second term is negative both of the factors must be negative. Thus, the factored equation is $(2x-1)(x-1)=0$. Find the solutions by making each factor equal to zero. $2x-1=0$; $x=\frac{1}{2}$ and $x-1=0$; $x=1$.

Question No.	Answer	Detailed Explanations
5	C	This problem can be easily solved by factoring out the greatest common factor 4x and then solving for each factor as shown: $4x^2-8x=0$ $4x(x-2)=0$ $4x=0$ and $x-2=0$ $x=0$ and $x=2$
6	E	All other aspects of man are mentioned in the poem.
7	A	The correct answer is A. The tone of this poem expresses a sense of disgust with mankind. The infant is described as "mewling" and the old man has no teeth, eyes, etc. The stages indicate a reliance on others, or dependence, which the narrator feels is sad.
8	B	The correct answer is B because you can see the round belly and the formal suit. The character as depicted seems to have an air of importance.
9	B	It seems almost circus-like as evident by the monkey, jester, children walking around on stilts, etc.
10	C	The poem unfolds like a story by first describing our "entrances and exits" and then goes on to describe the parts we play in this story.

Question No.	Answer	Detailed Explanations
1	B	We will write the equation in slope-intercept form, y=mx+b, which gives us the amount of water in the bucket after washing x cars. The question states that Alberto starts with 8 liters of water, which is equivalent to a y-intercept point (0,8).This means that we now have y=mx+8. The question also states that Alberto uses 3 liters of water for every car he washes. This means that the amount of water in the bucket decreases by 3 every time Alberto washes a car, which is the slope in the equation, but the slope is negative because the equation represents how many liters of water Alberto has after washing x cars. Now, we will enter the slope into the equation which gives us: y=mx+8;y=−3x+8. The slope represents the change in the number of liters of water Alberto has after each car. It is negative because the number is decreasing.
2	D	We will write the equation in slope-intercept form, y=mx+b, which gives the height of the tree after x years. The question states that the tree starts at 8 feet tall, which is equivalent to a y−intercept point (0,8). This means that we now have y=mx+8. The question also states that after three years, the tree is 17 ft tall, which is equivalent to the point (3,17). Now, we will use the point and substitute the x and y from the point into the equation to find the slope: y=mx+8; 17=m(3)+8; 9=3m; m=3.The equation is y=3x+8, with a slope of 3. The slope represents the number of feet the tree grows every year. It is positive because the height of the tree is increasing.
3	C	We will write the equation in slope-intercept form, y=mx+b, which will give us the length of the fence after Mike works on it for x hours. The question states that Mike has 8 meters of fence already completed, which is equivalent to a y−intercept point (0,8). This means that we now have y=mx+8. The question also states that Mike can complete 4 meters of fence per hour, which gives us a rate or slope. Now, we use these values to write the equation: y=mx+8;y=4x+8.The equation is y=4x+8, with a slope of 4. The slope represents the number of meters added to the length of the fence each hour. Since Mike works on the fence for 5 hours, we substitute 5 for x to find the length of the fence: y=4x+8;y=4(5)+8;y=20+8;y=28.

Question No.	Answer	Detailed Explanations
4	A	The equation above is already in slope intercept form of y=mx+b where m is the slope and b is the y intercept. Therefore the slope will be 4.
5	C	Rewrite the given equation in slope intercept form y=mx+b. In this case, m must equal 1 and b must be zero (0) in order to keep the function equivalent to the given equation. y=1x+0 Slope (m) will be 1 and the y-intercept (b) will be zero.
6	C	Answer C- based on the content and the intended audience, C is the best choice.
7	D	Answer D- all of these factors play an important part in choosing font and type.
8	A	Answer A- always research and verify your sources before using them.
9	B	Answer B- since this is a TV show, a video clip from the show would best accompany this.
10	C	The correct answer is C, because plagiarism is defined as stealing from a source without giving that source credit.

Question No.	Answer	Detailed Explanations
1	B	The path of the drone is that of a parabola opening upward. Thus, the lowest point on the graph, (4, 1) contains the minimum height of the drone. The minimum height is 100 feet because the y-scale is marked at ×100 feet (each unit on the y-axis represents 100 feet).
2	C	The table shows customer complaints and compliments per shift. The shift labeled "Noon to 6pm" received the next to the lowest complaints but the most compliments. Based on this information, we can conclude that this shift is doing the best job.
3	B	The slope of a function is its rate of change. When the slope is positive, the rate is increasing; when the slope is negative, the rate is decreasing. The period when the download speed is increasing the fastest is the interval where the graph has the steepest positive slope. This occurs in the interval 3 : 00 - 4: 00.
4	C	Since the y-intercept occurs at (0, 6) we know that it has to be answer choice A or C because the c value in standard form $ax^2 + bx + c$ is the y-intercept. Therefore the equation must have a "+6" at the end. You can use the x-intercepts and work backwards to find the rest of the equation.
		Since the zeros are -2 and -3, we can write factors (x+2)(x+3). Now use the FOIL method and multiply those together and arrive at $x^2 + 5x + 6$ which is answer choice C.
5	A	To find the axis of symmetry for a quadratic function, use the formula $x=-\frac{b}{2a}$ as shown below. $x=-\frac{b}{2a}=-\frac{18}{2(3)}=-(\frac{18}{6})=-3$
6	C	Answer C. Although A suggests equality, this text is written for men to be made equal as a result of the release from England's control over the US colonies.
7	D	Answer D- all appeals are used at various points in this document.

Question No.	Answer	Detailed Explanations
8	B	The tone of the Declaration of Independence is rebellious. Thomas Jefferson is summarizing all of the wrongdoing and explaining how the colonists should fight for what they want.
9	A	Answer A- Based on the evidence in the passage, the U.S. wants to ensure stability for its neighbors.
10	B	Answer B- in evaluating these statements, it is evident he is responding to critical comment of his military presence in other nations.

Question No.	Answer	Detailed Explanations
1	C	A function is positive when its graph is above the x-axis. Notice in the graph that the function's graph is above the x-axis in the interval(−3,2)U(5,∞) and below the x-axis on the interval (−∞,−3) U(2,5).
2	B	A function has a zero when the graph of the function touches or crosses the x-axis. Notice that the graph crosses the x-axis at x=-3,-1,1,3 which means the function has 4 zeros.
3	B	A polynomial function always has one fewer turning points that the degree of the function. The question states that the graph is that of a quintic polynomial, which is a 5th degree polynomial. Therefore, the graph would automatically have one less than five turning points, or four.
4	C	An even function's end behavior is the same as x −∞ or as x ∞. An odd function's end behavior is the opposite as x −∞ or as x ∞. Notice from the graph, that as x −∞,f(x) ∞ and as x ∞,f(x) −∞. Therefore, the function is an odd function.
5	B	The degree of a polynomial function can be determined by how many times the graph of the function crosses the x-axis, unless it bounces off the axis. The graph of g(x) crosses the x-axis five times. Therefore, g(x) is a 5th degree function. The degree of a polynomial function can also be determined by the number of turning points that the graph of the function has. The degree is always one more than the number of turning points. The graph has four turning points. Therefore, the function is a 5th degree function.
6	C	The answer is C. Kevin's struggle to stay awake the next day after staying up all night is a cause and effect relationship. Additionally, as a transition, does not accurately convey the cause and effect relationship the way that "consequently" does.
7	C	The answer is C. Using only one sentence structure over and over again creates writing that feels and sounds monotonous to the reader, producing boredom, even if the content is good.
8	D	The answer is D. A semicolon is used to combine two closely related sentences into one sentence.
9	C	The answer is C because the author's last name and then first name are listed first in the Works Cited entry.
10	A	Answer A because gasoline is stored in a tank just as money is stored in a vault.

Question No.	Answer	Detailed Explanations
1	D	The volume of the hopper is the difference between the full volume of the full pyramid and the volume of the truncated region. The width of the base of the truncated region is found by proportion as $(\frac{3m}{15m}) \times 5$ m = 1 m. $V=(\frac{1}{3}Bh)$pyramid--$(\frac{1}{3}Bh)$truncated $V=(\frac{1}{3})(5m \times 5m)(15m)-(\frac{1}{3})(1m \times 1m)(3m)$ $V=124m^3$
2	A	The fraction of the volume occupied by the volume is equal to the ratio of the volume of the liquid to the volume of the cone. The cone formed by the liquid has radius r and height h by proportion. Fraction = $\frac{\frac{1}{3}\pi r^2 h}{\frac{1}{3}\pi R^2 H}$ where R = radius of cone = 2r and H = height of cone = 2h. Fraction = $\frac{\frac{1}{3}\pi r^2 h}{\frac{1}{3}\pi (2r)^2(2h)}$ Fraction = $\frac{1}{8}$
3	C	The volume of the cone is $\frac{1}{3}\pi r^2 h$. The volume of the cone will be equal to the cylinder of fluid formed when poured into the can. The cylinder of fluid has volume $\pi r^2 h'$ where h' is the depth of the fluid. Therefore: $\frac{1}{3}\pi r^2 h=\pi r^2 h'$ $h'=\frac{1}{3}h$ So, $\frac{h'}{h}=\frac{1}{3}$.
4	C	Use the formula for volume of a cylinder: $V=\pi r^2 h$. In this case the radius (r) can be found by dividing the diameter in half. If the diameter is 2 meters then the radius will be 1 meter $(\frac{2m}{2}=1m)$. The height (h) is given as 3 meters. This information can now be substituted into the formula. $V=\pi r^2 h$ $V=\pi(1m)^2 3m=3.14(1m^2)(3m)=9.42m^3$.

Question No.	Answer	Detailed Explanations
5	B	Since the assumption is made in the problem that the snowball is a perfect sphere the formula for the volume of a sphere can be used: $V=\pi r^3$. The radius (r) will be ½ the diameter or($5in^2=2.5in$). Substituting this value into the formula: $V=\frac{4}{3}\pi r^3=\frac{4}{3}(3.14)(2.5in)^3$ $=\frac{4}{3}(3.14)(15.625in^3)=65.42in^3$ Of the possible choices $65in^3$ is closest to $65.42in^3$.
6	A	The correct answer is A because it encompasses the whole of the article; the other answer choices are the supporting details
7	A	One can assume it means large and powerful because the enemy appears so great judging by the urgency of his tone and the lengths he is going to to persuade the people to fight.
8	B	The answer is B, because that is where the rising action peaks and then starts to fall. Usually the moment of greatest emotional intensity in the story.
9	C	The answer is C, because you would persuade them to do something about the waste.
10	C	The answer is C, because the writer is using hyperbole to get you to visit Costa Rica. Words like wonderful and amazing are a red flag to the reader that you are being persuaded.

Question No.	Answer	Detailed Explanations
1	B	The question asks you to find the solutions to the quadratic equation $4x^2+8x-5=0$. Find the solutions using factoring. In factoring, you factor the first and the third term in such a way that the two factors add to give the middle term. Since the third term is negative, the two factors of that term must have different signs. Since the second term is positive, the larger of the factors must be positive. Lastly, since the middle term is even, the factors of the first term must both be even. Thus, the factored equation is $(2x-1)(2x+5)=0$. Find the solutions by making each factor equal to zero. $2x-1=0$; $x=½$ and $2x+5=0$; $x=\frac{-5}{2}$.
2	C	The question asks you to find the solutions to the quadratic equation $4x^2+8x+4=0$. Find the solutions using factoring. In factoring, you factor the first and the third term in such a way that the two factors add to give the middle term. Since all of the terms can be divided by 4, simplify the quadratic equation to $x^2+2x+1=0$. Also, since all of the terms are positive, the two factors of that term must have a positive sign. Lastly, notice that the quadratic trinomial is a perfect square trinomial, in that the first and third terms are squares and the middle term is twice the product of the factors of the first and third terms. Thus, the factored equation is $(x+1)^2=0$. Find the solutions by making the factor equal to zero. $x+1=0$; $x=-1$
3	D	The question asks you to find the solutions to the quadratic equation $4x^2-12x+9=0$. Find the solutions using factoring. In factoring, you factor the first and the third term in such a way that the two factors add to give the middle term. Since the third term is positive and the middle term is negative, the two factors must have a negative sign. Lastly, notice that the quadratic trinomial is a perfect square trinomial, in that the first and third terms are squares and the middle term is twice the product of the factors of the first and third terms. Thus, the factored equation is $(2x-3)^2=0$. Find the solutions by making the factor equal to zero. $2x-3=0$; $x=\frac{3}{2}$

Question No.	Answer	Detailed Explanations
4	A	The question asks you to find the solutions to the quadratic equation $x^2+19x+60=0$. Find the solutions using factoring. In factoring, you factor the first and the third term in such a way that the two factors add to give the middle term. Since all of the terms are positive, the two factors of that term must have positive signs. Thus, the factored equation is $(x+4)(x+15)=0$. Find the solutions by making each factor equal to zero. $x+4=0$; $x=-4$ and $x+15=0$; $x=-15$.
5	D	This problem can be easily solved by rearranging the equation so that it is solved for zero and then factoring as shown: $x^2+10x=-25$ $x^2+10x+25=0$ $(x+5)(x+5)=0$ Since both factors are exactly the same, you will only have one solution to this problem. $x+5=0$ $x=-5$
6	D	The correct answer is D because she is not interested in her sister's book since it has no conversations in it.
7	B	The correct answer is B. Alice found herself falling down a very deep well. The following lines from Paragraph 6 clearly explains that she was curious. 'she had plenty of time as she went down to look about her and to wonder what was going to happen next. First, she tried to look down and make out what she was coming to, but it was too dark to see anything; then she looked at the sides of the well, and noticed that they were filled with cupboards and book-shelves; here and there she saw maps and pictures hung upon pegs. She took down a jar from one of the shelves as she passed; it was labeled 'ORANGE MARMALADE', but to her great disappointment it was empty: she did not like to drop the jar for fear of killing somebody, so managed to put it into one of the cupboards as she fell past it'.
8	B	The correct answer is B. The quote is endearing and the reader can really see into the childlike mind of Alice.
9	B	The correct answer is B because she starts to compare herself to the lens of the telescope and a comparison between two things that uses "like" or "as" is called a simile.
10	D	The correct answer is D. The imagery here allows the reader to imagine Alice like a telescope opening and shutting to view things at a distance. She wants to be like a telescope because old-fashioned telescopes can decrease in size. She wants to decrease in size enough to fit through the door.

Question No.	Answer	Detailed Explanations
1	A	Using the complex number system, we can simplify $\sqrt{144} + \sqrt{-121}$. The first step is to change the radical expression to $\sqrt{144} + \sqrt{-1} \cdot \sqrt{121}$. Now, we know that $\sqrt{-1} = i$. Therefore, $\sqrt{144} + \sqrt{-1} \cdot \sqrt{121} = 12 + 11i$.
2	C	We know that $\sqrt{-1} = i$. Therefore, we can simplify the radicals: $\sqrt{25} \cdot \sqrt{-36} = \sqrt{25} \cdot \sqrt{-1} \cdot \sqrt{36}$. Next, simplify each radical and multiply: $\sqrt{25} \cdot \sqrt{-1} \cdot \sqrt{36} = 5 \cdot 6i = 30i$.
3	C	We know that $i^2 = -1$. Then , $i^3 = i^2 . i = -1.i = -i$ and $i^4 = i^2.i^2 = (-1)(-1) = 1$. Therefore, we find how many i^4 are in i^{451} : $451 = 4(112)+3$. Now using exponent rules : $i^{451} = i^{448}. i^3$. Since $i^4= 1$, $(i^4)^{112} = 1^{112} = 1$. Then , $i^{451}= 1.i^3 = -i$.
4	A	Using the complex number system, we can simplify $\sqrt{-256} + \sqrt{225}$ The first step is the change the radical expression to $\sqrt{-1} . \sqrt{256} + \sqrt{225}$ Now, we know that $\sqrt{-1} =i$. Therefore, $\sqrt{-1}.\sqrt{256} + \sqrt{225} = 16i+15=15+16i$
5	D	This problem has a real part (3) and an imaginary part (7i). You cannot add real and imaginary numbers together since they are not like terms. You can only add/subtract real numbers to/from other real numbers. Likewise you can only add/subtract imaginary numbers to/from other imaginary numbers. Therefore the answer remains 3+7i.
6	A	The correct answer is A. When you cite your sources correctly you inform the reader of where you gathered the information from.
7	D	The correct answer is D, because these are the defining characteristics of a good research statement.
8	D	The correct answer is D, because those are all different styles.
9	A	The correct answer is A, because you would place your source in a separate section of your paper.
10	B	The correct answer is B, because you should place all direct quotes in quotation marks, unless the quote is longer than 4 lines. In that case, you would follow a different pattern

Question No.	Answer	Detailed Explanations
1	B	We know that √-1 = i. Therefore, we can simplify radicals: √-25.√-36=√-1.√25.√-1√36. Next simplify each radical and multiply: √-1.√25.√-1√36=5i.6i=30i^2=-30
2	A	Using the complex number system, we can simplify √-256+ √-225. The first step is to the change the radical expression to √-1. √ 256+ √-1. √225. Now, we know that √-1=i. Therefore, √-1. √256+ √-1. √225=16i+15i=31i
3	C	The question asks us to multiply (2+3i).(4+i). These two expressions are both complex numbers. We will perform this multiplication using the FOIL process. Thus, (2+3i).(4+i)=2(4)+2(i)+3i(4)+3i(i)=8+2i+12i+3i^2. Now, since i^2=-1, we will substitute in the previous expression: 8+2i+12i+3i^2=8+2i+12i+3(-1). Combine like terms and we have the final answer: 8+2i+12i+3(-1)=8+2i+12i-3=5+14i
4	D	The question asks us to multiply (-5+3i).(-4-9i). These two expressions are both complex numbers. We will perform this multiplication using the FOIL process. Thus, (-5+3i).(-4-9i)=-5(-4)+(-5)(-9i)+(3i)(-4)+(3i)(-9i) and -5(-4)+(-5)(-9i)+(3i)(-4)+(3i)(-9i)=20+45i-12i-27i^2. Now, since i^2=-1,we will substitute in the previous expression: 20+45i-12i-27i^2=20+45i-12i-27(-1). Combine like terms and we have the final answer: 20+45i-12i-27(-1)=20+45i-12i+27=47+33i
5	A	$(6i)^2=6^2*i^2=36*-1=-36$ Use product to a power rule for exponents to see that the 6 and the i will both need to be squared. Six squared is 36 (6*6). Then i^2=−1 by definition. Get the final result by multiplying the two together: 36x-1=-36
6	B	Answer B is the correct choice because the article discusses how to end the stigma surrounding mental health.
7	C	Answer C is the correct choice because spreading awareness and educating the public about mental health conditions is encouraged in the passage.
8	B	Answer B is the correct choice because spreading information and having conversations about mental health lessens the stigma.
9	C	Answer C is the correct choice because when famous people share their stories about mental health others can join in without fear of judgement.
10	A	Answer A- evidence from the text supports a fair trial.

Question No.	Answer	Detailed Explanations
1	C	The question states that one dollar is invested in an account that accumulates 20% interest per year. This means that the value of the account increases proportionally based on the amount in the account already. The formula for this type of function is $f(x)=1(1.2)^x$ which has a growth factor of 1.2 per year. A graph of this function is below. 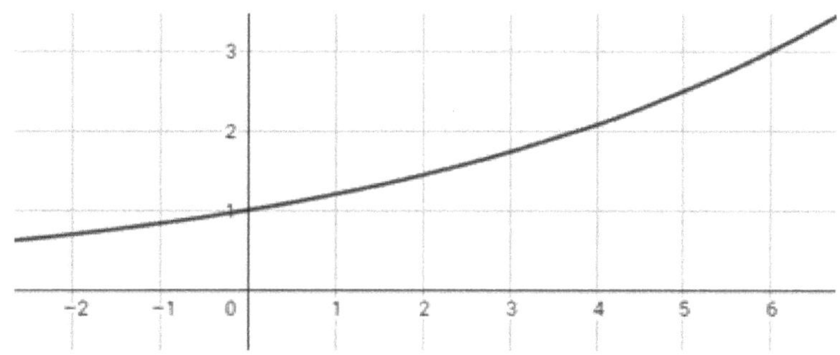
2	C	The question states that the pressure inside a chamber of a compressed oxygen tank begins at -2 PSI and every day, the pressure decreases 20% more than the previous day. This means that the pressure in the chamber decreases proportionally based on the previous day's pressure. The formula for this type of function is $f(x)=-2(1.2)^x$ which has a growth factor of 1.2 per day, and even though it is in the negative direction, the function is an exponential growth function. A graph of this function is below. 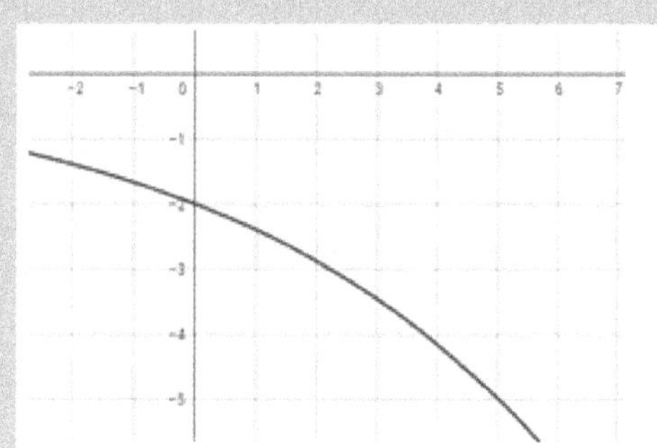

Question No.	Answer	Detailed Explanations
3	C	The question states that $200 was invested in an account that accumulates a guaranteed 1% every six months. The formula for this type of function is $f(x)=200(1+(\frac{0.1}{2}))^{2x}$ which has a growth factor of 1.005 every six months. The function is an exponential growth function. A graph of this function is below. 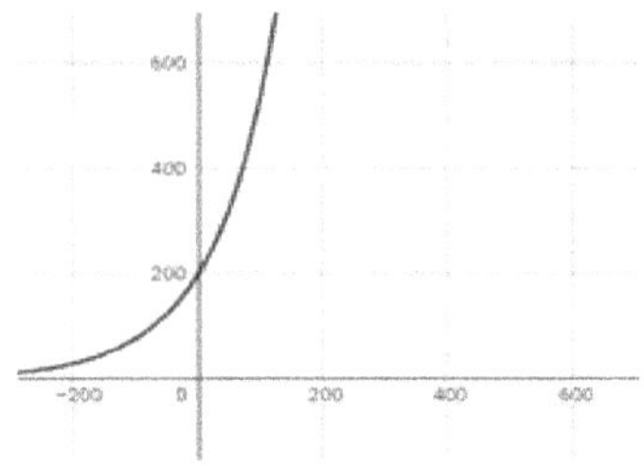
4	A	The question states that an elderly man has $40,000 in an account and he has to withdraw a specific amount every year to ensure the account is empty in 10 years. This information equates to a y-intercept of 40,000. Then if he withdraws the money so it is gone in 10 years, this equates to a slope of -4000. This situation can be represented by the linear function f(x)=-4x+40 (where the units is times 1,000.) A graph of this function is below. 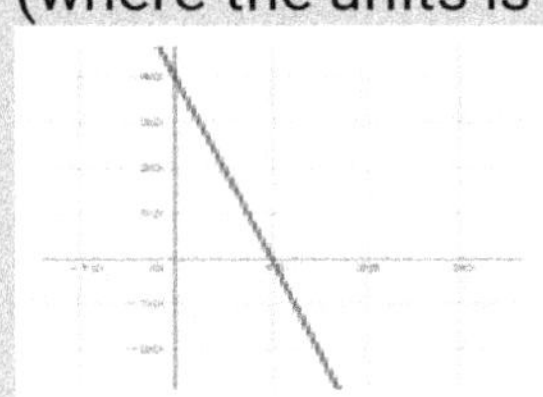
5	C	Since the dog food plant is producing a steady rate of 20 tons per month, a linear function is the best representation of this situation. The linear function in these choices is C so this is the best function to represent this situation.
6	D	Based on the context contrasting a wunderkind with someone achieving success when they are older, the correct answer is D.
7	A	The answer is A because disseminate means to spread or disperse
8	B	The answer is B because dissidents means protesters or dissenters.
9	A	The answer is A. The context of the sentence tells the reader that unprecedented can refer to an event that has never occurred before.
10	C	The answer is C because 'facilitate' means to make it easier or to enable.

Question No.	Answer	Detailed Explanations
1	A	Rules: Equation of a Circle:$(x-h)^2+(y-k)^2=r^2$; C(h,k); r=radius. Using the formula for the equation of a circle above. Substitute the h and k values of the center into the equation and the radius. Don't forget to square the radius.
2	D	Rules: Equation of a Circle: $(x-h)^2+(y-k)^2=r^2$; C(h,k); r=radius. A circle in standard position is centered at (0, 0). Since this circle has its center at (2, -3) that means it has shifted right 2 units and down three units. The radius = $\sqrt{16}=4$.
3	A	Rules: Equation of a Circle: $(x-h)^2+(y-k)^2=r^2$; C(h,k); r=radius. The formula for the equation of a circle below shows the coordinates of the center (h, k). Since h = 5 and k = -3, the center is (5, -3). To find the diameter, we need to find the radius first since it is included in the formula and then double it to get the diameter. We can find $r=\sqrt{36}=6$. So the diameter is 12.
4	C	The question asks for the equation in the form $(x-h)^2+(y-k)^2=r^2$. The general form of the point that is the center of a circle is (h,k), so h=1, and k=−3. This gives us the equation $(x-1)^2+(y+3)^2=r^2$. The question says the center of the circle is at the point (1,−3) and the point (4,0) is on the circle. This means that the point on the circle is 3 units to the right and 3 units up from the center. Find the radius squared using the Pythagorean Theorem, $x^2+y^2=r^2$, so $r^2=3^2+3^2=18$. The equation of the circle is $(x-1)^2+(y+3)^2=18$.
5	C	The question asks for the equation in the form $(x-h)^2+(y-k)^2=r^2$. The general form of the point that is the center of a circle is (h,k), so h=1, and k=−4. This gives us the equation $(x-1)^2+(y+4)^2=r^2$. The question says the center of the circle is at the point (1,−4) and the point (6,4) is on the circle. This means that the point on the circle is 5 units to the right and 8 units up from the center. Find the radius squared using the Pythagorean Theorem, $x^2+y^2=r^2$, so $r^2=5^2+8^2=89$. The equation of the circle is $(x-1)^2+(y+4)^2=89$.
6	B	Answer B. The Constitution established the rules, but the purpose of the Preamble was to justify the need for the Constitution.

Question No.	Answer	Detailed Explanations
7	B	Answer is B. This text is to inform people of the rules they will be establishing.
8	D	The answer is D because the passage both persuades the reader of the benefits of a later school start time and supports that claim with general factual information.
9	D	The answer is D because it captures the main idea of the entire passage.
10	A	The answer is A. Based on the context of expecting severe punishment, clemency means mercy.

Question No.	Answer	Detailed Explanations
1	B	Convert each equation to slope-intercept form. The system becomes: y=x−3; y=7x+3. For each equation, plot the y-intercept point, and use the slope to find another point on the line. Then, identify the point where the two lines intersect. The graph of this system of equations is below. 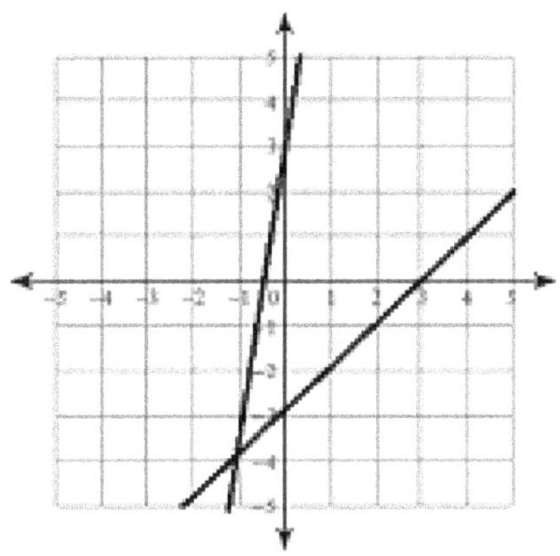 Notice that the two lines intersect at the point (−1,−4).
2	A	Convert each equation to slope-intercept form. The system becomes: y=-2x + 1; y=x - 5. For each equation, plot the y-intercept point, and use the slope to find another point on the line. Then, identify the point where the two lines intersect. The graph of this system of equations is below 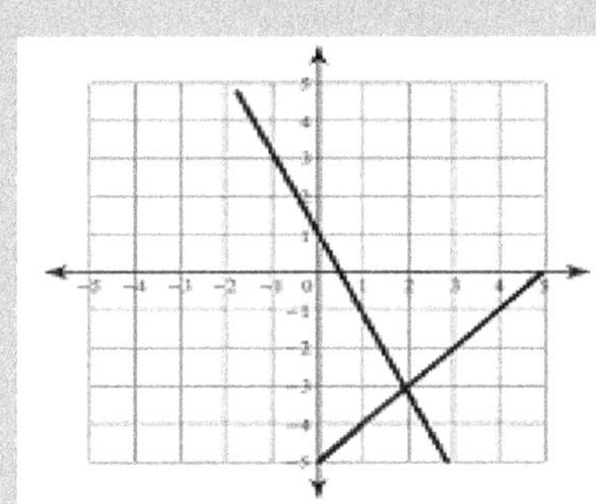 Notice that the two lines intersect at the point (2,−3).

Question No.	Answer	Detailed Explanations
3	D	Convert each equation to slope-intercept form. The system becomes y = 3x - 4 and y = - ½x + 3. For each equation, plot the y-intercept, then use the slope to find another point on the line. Then, identify the point where the two lines intersect. The graph of this system of equations is below. 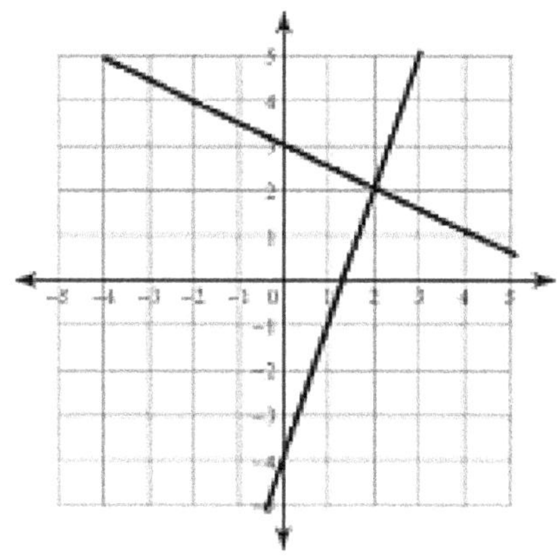 Notice that the two lines intersect at the point (2,2).
4	B	Convert the first equation to slope-intercept form. The system becomes y = x + 2 and x = -3. For each equation, plot the y-intercept point, and use the slope to find another point on the line. Then, identify the point where the two lines intersect. The graph of this system of equations is below. 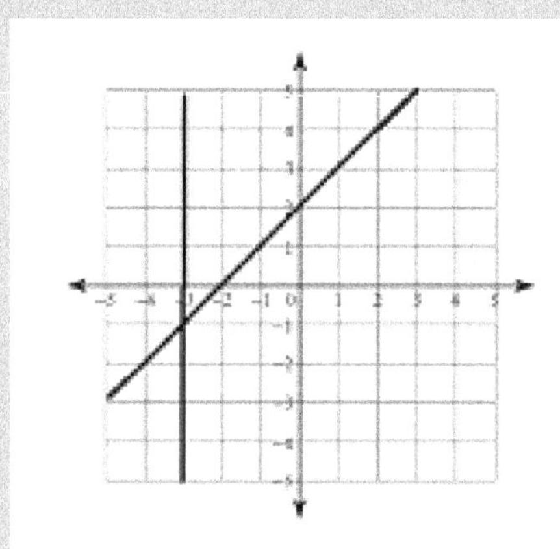 Notice that the two lines intersect at the point (−3,−1).

Question No.	Answer	Detailed Explanations
5	B	We could let x represent the first number and y represent the second number. Then the two equations would be as follows $x+y=24$; $x-y=6$ Using the elimination (addition) method, you could eliminate the y's and solve for x. $x+y=24$ $x-y=6$ $2x=30$ $x=15$ Now use the fact that $x=15$ and plug that value into one of the original equations to find y. $x+y=24$ $15+y=24$ $y=9$ So the intersection point and final solution to the system of equations is (15, 9).
6	B	The answer is B. The narrator questions why the person to whom he is telling his story would think him insane. He believes that the carefulness he exhibited in executing his plan to murder the old man clearly demonstrates his sanity.
7	D	The answer is D. Rather than the police discovering the narrator's crime, the narrator's guilt, exacerbated by the sound in his mind of the old man's beating heart, drives him to confess his crime. The guilt was too overwhelming for him to hold the truth any longer.
8	B	The answer is B. The narrator repeatedly points to the patience and caution he exhibited in his planning and execution of the murder of the old man.
9	A	The answer is A. As the narrator repeatedly proclaims his sanity while committing actions that might point to his insanity, the reader is led to wonder if the events the narrator is telling are true.
10	B	The answer is B. The context shows that "believe" is the only synonym that could replace "fancy," and the sentence communicates the same idea.

Question No.	Answer	Detailed Explanations
1	B	We know that $i^2=-1$. Then $i^3=i^2.i = -1.i=-i$, and $i^4 = i^2. i^2 = (-1)(-1) = 1$. Now, using exponent rules $i^{34}= i^{32}.i^2=(i^4)^8.i^2$ Since $i^4=1(i^4)^8=1^8=1$. Then, $i^{34}=1.i^2=1. -1=-1$
2	C	We know that √-1 = i Therefore, we can simplify the radicals: √49 . √-9 = √49 . √-1 . √9 = 7.i.3 = 7-3i
3	C	We know that $i^2= -1$. Then, $i^3 = i^2.i =-1.i=-i$,and $i^4 = i^2.i^2 =(-1)(-1) = 1$. Now , using exponent rules, we will multiply: $7.(-6)i^{137} = -42i^{137}$ Next, we find how many i^4 are in i^{137} : 137=4(34)+1 . Now, using exponent rules again : $-42i^{137} = -42i^{4.34+1}=-42(i^4)^{34}.i^1$ Since $i^4=1.(i^4)^{34} = 1^{34}=1$ .Then $-42i^{137} = -42.1.i =-42i$
4	A	There is no normal or "real" way to solve the problem √-1 since there is no number that you can square and get a negative answer. Imaginary numbers were invented as a way to solve these “un-solvable” problems.
5	D	Since an imaginary number is defined as i=√-1, we should be able to square both sides of that equation and get a value for i^2 like so, $(i)^2=(\sqrt{-1})^2$. That simplifies to i^2 on the left and -1 on the right. A square root and a square are opposite operations that eliminate one another, so the -1 remains.
6	A	The answer is A- this would be the most valuable and valid resource
7	B	The answer is B- any website with .edu is more valid than other sites and has research backing its findings.
8	A	The answer is A- the focus is the female leaders not the male connection to them.
9	B	The correct answer is B, because you should vary your sources so they include both print and non-print resources.
10	B	The correct answer is B, because you set up your question after brainstorming on the topic a bit.

Question No.	Answer	Detailed Explanations
1	C	We know that $i^2=-1$. Then, $i^3=i^2.i=-1.i=-i$, and $i^4=i^2.i^2=(-1)(-1)=1$.Now using exponent rules, we will multiply $7i^{127}.(-6)i^{138}=-42i^{127+138}=-42i^{265}$. Next, we find how many i^4 are in i^{265}:265=4(66)+1.Now, using exponent rules again:$-42i^{265}=-42i^{4.66+1}=-42(i^4)^{66}.i^1$ Since $i^4=1 (i^4)^{66}=1^{66}=1$.Then, $-42i^{265}=-42.1.i=-42i$
2	B	The question asks us to add (9+4i)+(-6-2i). These two expressions are two complex numbers. Even though $i^2 = -1$, we can treat the letter i as a variable when performing some operations. Therefore, we add the two complex numbers by combining like terms. Thus, (9 + 4i) + (-6 - 2i) = (9 + (-6)) + (4i + (-2i)) = 3 + 2i.
3	B	The question asks us to multiply (6-7i).(4-5i). These two expressions are both complex numbers. We will perform this multiplication using the FOIL process. Thus, (6-7i).(4-5i)=6(4)-6(-5i)+(-7i)(4)+(-7i)(-5i) and 6(4)+6(-5i)+(-7i)(4)-(-7i)(-5i)=$24-30i-28i+35i^2$. Now, since $i^2=-1$, we will substitute in the previous expression: $24-30i-28i+35i^2=24-30i-28i-35$. Combine like terms and we have the final answer:24-30i-28i-35=-11-58i.
4	C	Use the distributive property and multiply both the real and the imaginary part of 2−3i times 5: 5(2−3i)=5*2−5*3i=10−15i. The resulting answer is 10−15i.
5	D	$(2+4i)(1-i)=2-2i+4i-4i^2=2+2i+4=6+2i$ Use the FOIL method for multiplying two binomials together, as shown above. Simplify by combining the two middle terms(2i+4i) since they are like terms. Next, simplify $-4i^2$ to be $-4 \cdot (-1) = 4$, since $i^2 = -1$ by definition. Finally we can add the 4 and 2 together to get the answer 6+2i.
6	D	The answer is D. All of the other words have negative connotations associated with them. The demonstration does not.

Question No.	Answer	Detailed Explanations
7	B	The answer is B, because hyperbole is an exaggeration used for emphasis.
8	D	The answer is D, because a metaphor draws a direct comparison between two objects; in this case, the author states that there is a dark cloud over the girl's head when describing her sadness.
9	C	The answer is C. All of the other words have a positive connotation.
10	D	The answer is D. A euphemism is a milder or more indirect word or expression which is substituted for a word which is considered to be too harsh or blunt while referring to something which is unpleasant or embarrassing.

Question No.	Answer	Detailed Explanations
1	A	The graph shows a linear function, so the formula, in slope-intercept form is f(x)=mx+b, where m is the slope of the line (rise over run) and b is the y-intercept. Looking at the graph, select the two points (0,4) and (4,5) to calculate the slope. The formula and calculations are: m= $\frac{(y_2-y_1)}{(x_2-x_1)} = \frac{(5-4)}{(4-0)} = \frac{1}{4}$ The graph shows that the y-intercept is (0,4), so b = 4. The function whose graph is shown is $f(x)=(\frac{1}{4})x+4$.
2	A	The questions states that the sequence is arithmetic. The formula for an arithmetic sequence is $an=a_1+d(n-1)$,where a_n is the n_{th} term, a_1 is the first term, and d is the difference between each term. The table shows that $a_1=5$ and each term is 5 more than the previous term so d=5.Using these numbers, the sequence rule is an=5+5(n−1)=5+5n−5=5n.
3	B	The graph shows that the function is an exponential growth function. The formula for an exponential function is $f(x)=ab^x$, where a is the y-intercept and b is the growth factor. If the exponential function is a growth function, then b>1. If the exponential function is a decay function, then 0<b<1. The graph passes through the points (0,5) and (1,2), which are one unit apart on the x-axis, so find b by dividing 2 by 5: b = 2/5 = 0.4. The function whose graph is shown in the question is $f(x)=5(0.4)^x$.
4	A	The y-intercept is -3 so there should be a - 3 on the end of the equation, and the only one that has this is A. Choice B would cross the y-axis at +3 and choices C and D would cross the y-axis at 0.
5	B	The only one of these that would be 20 for 0 years is choice B. This function also works for x=1 and x=2.
6	A	The correct answer is A. The purpose of this speech is to respectfully dedicate a portion of the ground for the soldiers who have died in battle. It is mentioned in the passage - "We have come to dedicate a portion of that field, as a final resting-place for those who here gave their lives".
7	D	The answer is D because he is very respectful of the soldiers who gave their lives, but he is also sounds a little worried about the future of the country. He questions whether or not it will even survive this war.

Question No.	Answer	Detailed Explanations
8	B	Metaphor is used in the phrase "in your hands." Allusion- reference to the Constitution. Call to action- clear evidence that people need to get involved to make change.
9	A	Lincoln's purpose is to unite everyone for a common goal: ending the Civil War.
10	D	Answer D- this shows how the balance of power among the states is based on all three plans, culminating in the Connecticut Plan.

Question No.	Answer	Detailed Explanations
1	C	Rules: Equation of a Circle: $(x-h)^2+(y-k)^2=r^2$; C(h,k); r=radius. Use the formula for the equation of a circle above. Therefore, h = -1, k = 7, and r =√23. The coordinates for the center of a circle are (h, k) so (-1, 7) is the center of this circle and the radius =√23.
2	B	Rules: Equation of a Circle: $(x-h)^2+(y-k)^2=r^2$; C(h,k); r=radius. The values in the formula for the equation of a circle are h, k, and r. The h and k values are the coordinates of the center. We must have r for the radius, so we will need to take half of the diameter to get r = 5. Don't forget to square the radius when it is in the formula.
3	C	Rules: Equation of a Circle: $(x-h)^2+(y-k)^2=r^2$; C(h,k); r=radius. The original circle, $x^2+y^2=49$, has its center located at (0, 0) since the h and k values in the equation are both zero. Shifting left 12 will make h = -12. Shifting down 9 will make k = -9. Plug those values into the formula for the equation of a circle above. The radius did not change.
4	C	The question asks for the equation in the form $(x-h)^2+(y-k)^2=r^2$. The general form of the point that is the center of a circle is (h,k), so h=2, and k=−1. This gives us the equation $(x-2)^2+(y+1)^2=r^2$. The question says the center of the circle is at the point (2,−1) and the point (4,2) is on the circle. This means that the point on the circle is 2 units to the right and 3 units up from the center. Find the radius squared using the Pythagorean Theorem, $x^2+y^2=r^2$, so $r^2=2^2+3^2=13$. The equation of the circle is $(x-2)^2+(y+1)^2=13$.
5	A	The question asks for the equation in the form $(x-h)^2+(y-k)^2=r^2$. The general form of the point that is the center of a circle is (h,k), so h=0, and k=3. This gives us the equation $x^2+(y-3)^2=r^2$. The question says if the center of the circle is at the point (0,3) and the point (3,7) is on the circle. This means that the point on the circle is 3 units to the right and 4 units up from the center. Find the radius squared using the Pythagorean Theorem, $x^2+y^2=r^2$,so $r^2=3^2+4^2=25$. The equation of the circle is $x^2+(y-3)^2=25$.
6	A	The correct answer is A because the other three deal with other aspects of the research process.
7	D	The correct answer is 'none of the above' because all of those things must be included in your bibliography or works cited.
8	A	The correct answer is A because those are the most valid types of sites.
9	B	The answer is B because the color and sound words appeal to the reader's sense of sight and sound.
10	C	The answer is C because it summarizes the most important idea conveyed throughout the entire passage.

Question No.	Answer	Detailed Explanations
1	C	The question asks you to solve this system of equations algebraically. $y=-2$; $x^2+y^2=16$ Substitute −2 into the second equation for y and solve for x. $x^2+y^2=16$ $x^2+(-2)^2=16$ $x^2+4=16$ $x^2=12$ $x=\pm\sqrt{12}$ or $\pm2\sqrt{3}$. So $(\pm2\sqrt{3}, -2)$
2	B	The question asks you to solve this system of equations graphically. $y=x^2-8x+2$; $x+y=4$. The graph of the system of equations is below. The intersection points given below are approximated. 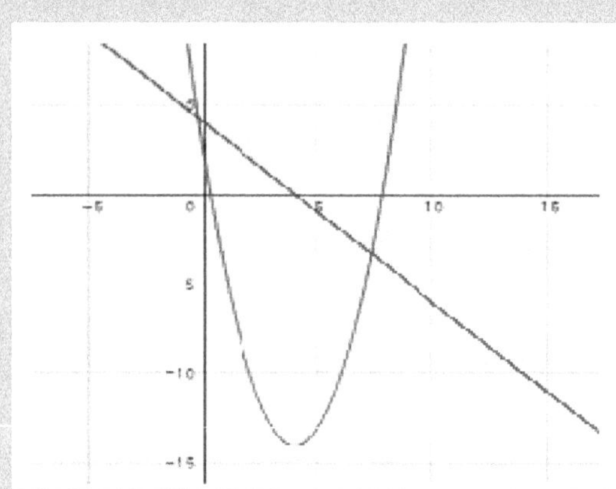 Notice that the intersection on the left is at the approximate point (−0.27, 4.27) and the intersection on the right is at the approximate point (7.27, −3.27).The answer is (−0.27, 4.27), (7.27, −3.27).
3	B	The question asks you to solve this system of equations algebraically. $x=2$; $x^2+y^2=25$. Substitute 2 into the second equation for x and solve for y. $x^2+y^2=25$ $2^2+y^2=25$ $4+y^2=25$ $y^2=21$ $y=\pm\sqrt{21}$ So $(x, \pm\sqrt{21})$

Question No.	Answer	Detailed Explanations
4	B	The question asks you to solve this system of a quadratic equation and a linear equation algebraically. $x^2+y^2=36$; $x=4$ Substitute 4 into the first equation for x and solve for y. $x^2+y^2=36$ $4^2+y^2=36$ $16+y^2=36$ $y^2=20$ $y=\pm\sqrt{20}$ $y=\pm2\sqrt{5}$ So $(x, \pm2\sqrt{5})$
5	B	The question asks you to solve this system of equations algebraically. $x^2+y^2=9$; $y=3$ Substitute 3 into the first equation for y and solve for x. $x^2+y^2=9$; $x^2+(3)^2=9$ $x^2+9=9$ $x^2=0$ $x=0$
6	B	The correct answer is B because the main issue in the story is that Richard Cory shoots himself at the end of the poem. His struggle is internal, as evident by how little the town understands him.
7	D	The correct answer is D because the rising action leads to this point when Richard Cory takes his life. All conflict leads to that surprise.
8	D	The correct answer is D because all those quotes seem to emphasize how superior he is. "We" people; suggests a separateness and "glittered when he walked" and then the direct characterization of him being "rich".
9	C	The answer is C because though Cory is rich, he is also unhappy.
10	D	The correct answer is D. The meat and the bread are simply symbols of the struggles of these townspeople; that they do without.

Question No.	Answer	Detailed Explanations
1	C	The question asks you to identify the point that is on the graph of $f(x)=\frac{4x}{5}-\frac{9}{5}$. Substitute the x-value in each point into the function to find the y-value. If you substitute $x=6, f(6)=(\frac{4}{5})(6)-(\frac{9}{5})=\frac{24}{5}-\frac{9}{5}=\frac{15}{5}=3$. The other choices do not produce the correct y-value.
2	A	The question asks you to identify the point that is on the graph of $f(x) = -19x + 45$. Substitute the x-value in each point into the function to find the y-value. If you substitute $x=9, f(9)=-19(9) + 45=-171 + 45=-126$. The other choices do not produce the correct y-value.
3	B	The question asks you to identify the point that is on the graph of $f(x)=\frac{3}{7}x + \frac{15}{7}$. Substitute the x- value in each point into the function to find the y-value. If you substitute $x=-3, f(-3)=(\frac{3}{7})(-3) + (\frac{15}{7})= -\frac{9}{7} + \frac{15}{7}=\frac{6}{7}$. The other choices do not produce the correct y-value.
4	B	The break-even point happens when profit is zero. This means Samantha has sold enough cookies to make back all the money she invested but hasn't made any "profit" yet. The break-even point would equal zero. We can find the x-value by plugging in a zero for y and solving for x like so: $y = 2x - 16$ $0 = 2x - 16$ $16 = 2x$ $x = 8$. This means after selling 8 dozen cookies, Samantha will have made back the money she spent on supplies and ingredients. She will need to sell more than 8 dozen cookies to start making a profit. You can see this on the graph also as shown below. 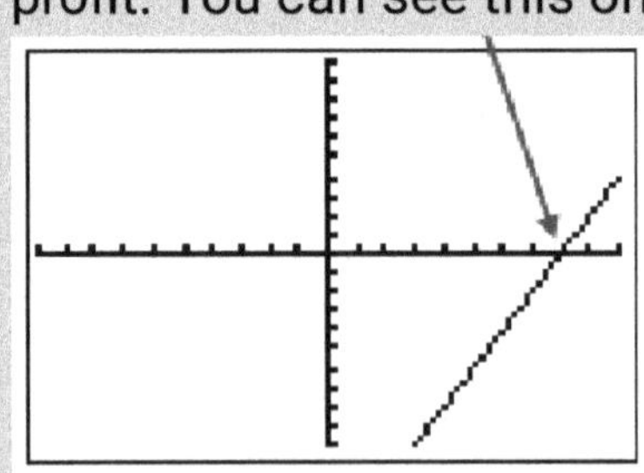

Question No.	Answer	Detailed Explanations
5	A	We can set the equation equal to zero and solve for x to find the break-even point, after which the company will start to turn a profit. $x^2 - 10 = 0$ $x^2 = 10$ $\sqrt{x^2}=\sqrt{10}$ $x = 3.16$ So the break-even happens somewhere between day 3 and day 4 and that is when the company will start to make a profit.
6	D	The answer is D- all of these are required
7	A	The answer is A- be cautious of sites with a .com or .org as anyone can purchase these.
8	C	The answer is C- always verify the credibility of the source.
9	C	The answer is C- .gov sites are government sites and would be the only example of an authoritative source.
10	D	The answer is D- all forms are correct.

Question No.	Answer	Detailed Explanations
1	B	At the origin, the aircraft is not moving. Since the drag is equal to a coefficient times the velocity, and the airplane is not moving, the velocity is zero. Thus, any coefficient times zero is zero. If the airplane is not moving, there is no drag.
2	B	The graph does not pass through the origin because a male elephant is 83.5 cm tall at birth.
3	A	Based on the information given, the driver was able to drive up to 70 miles per hour when not in a city. The graph shows that he/she did maintain that speed for about two hours. This speed equates to 7 grids along the y-axis. Therefore each mark (grid) represents 10 miles per hour.
4	D	The value of the graph at the origin is 0 for the person's elevation and 0 for the distance to the horizon. This means that if the person's line of sight is at the same level as the ground, the horizon is at the person's location.
5	C	The graph has the shape of a logarithmic function, which increases but gradually levels off. It passes through the origin because log(1)=0, meaning 0 carat corresponds to 0 diamonds purchased. Logarithmic graphs rise rapidly at first and then grow more slowly as x increases. Therefore, the correct description is that the quantity increases quickly at first, then more slowly.
6	C	The answer is C because there is no limit to the number of sources you can have in a Works Cited.
7	D	The answer is D because the official site of the science museum will offer the most accurate, detailed, and up-to-date information needed for planning the field trip.
8	A	The answer is A because it follows the rules for proper MLA format.
9	B	The answer is B because the statistics will be numerically accurate and they are from a government agency.
10	D	The answer is D because it follows the rules of proper MLA formatting.

Question No.	Answer	Detailed Explanations
1	B	The number of widgets in the warehouse is calculated by subtracting the number of widgets sold from the beginning inventory of widgets. The problem states that the warehouse has 950 widgets when the business opened. The number of widgets sold is calculated by multiplying the sales per week by the number of weeks. The problem also states that the business expects to sell 45 widgets every week and the variable w represents the number of weeks. Therefore, the function that represents the number of widgets in the warehouse after w weeks is $I(w)=950-45w$.
2	A	The number of hamburgers in the cooler is calculated by subtracting the number of hamburgers sold from the beginning inventory of hamburgers. The problem states that the restaurant has 1,090 hamburgers at the beginning of the week. The number of hamburgers sold is calculated by multiplying the sales per day by the number of days. The problem also states that the restaurant business expects to sell 150 hamburgers every day and the variable p represents the number of days. Therefore, the function that represents the number of hamburgers in the cooler after p days is $H(p)=1090-150p$.
3	B	The company has 150 sheets on hand and receives t truckloads with 80 sheets in each truckload. Hence, the total number of sheets the company has will be $P(t) = 150 + 80t$.
4	B	The number of students at the school is calculated by adding the number of students who were at the school in the beginning to the number of new students. The problem states that the school started with 60 students. The number of additional students is calculated by multiplying the number of new students per year by the number of years. The problem also states that the school expects to enroll 35 students every year and the variable x represents the number of years. Therefore, the function that represents the number of students at the school after x years is $S(x)=60+35x$.

Question No.	Answer	Detailed Explanations
5	C	Net profit is the result of subtracting the expenses from the total revenue. The problem says the theater spent \$495 to produce the play, so expenses total \$495. The revenue is the price of each ticket times the number of tickets sold, so the total revenue is \$15t. Therefore, the function that represents net profit is P(t) = 15t - 495.
6	D	The answer is D. A thesaurus lists words in alphabetical order and gives both the synonyms and antonyms.
7	D	The answer is D, because economy means the wealth and resources of a country.
8	B	The answer is B, because legislative means relating to laws or the making of them.
9	C	The answer is C. The lights of the yacht are being compared to fireflies vanishing into the night.
10	B	The answer is B, because concepts means abstract ideas or notions.

Question No.	Answer	Detailed Explanations
1	D	The perimeter of a figure is found by adding all of the sides together. Since a square has 4 congruent sides, we can multiply the length of the side, 15 ft, times 4 to get 60 ft.
2	B	Use the formula for the area of a triangle above. The legs of a right triangle are the base and the height. The area is 154.875 square units, which rounds to 155 when rounding to the nearest whole number.
3	C	To find the perimeter of a figure, we must add the length of each side. To find the missing side of the triangle, we need to use the Pythagorean Theorem and find the hypotenuse, c. The hypotenuse is $5\sqrt{5} \approx 11.2$. Then add all three sides together to get the perimeter.
4	A	The distance around a circle is called the circumference. Plug the radius, 6 ft, into the circumference formula above. C = 37.70 ft.
5	C	To find the area of a circle using the formula above, we need to know the radius of the circle. Since the radius is ½ the diameter, the radius is 2.5 cm. Plug 2.5 cm into the formula for area of a circle above. $A \approx 20$ cm^2.
6	D	The answer is D. Dramatic irony is when the audience knows something that a character does not know.
7	A	Answer is A, because the pepper is a non-human object being given human qualities – the ability to tickle.
8	A	Answer A because an oxymoron is formed when two opposite words are paired to create a new phrase.
9	B	The answer is B because “predetermined” means fixed or prearranged.
10	A	The answer is A because the text refers to the circle of life, the “road of Destiny”, and one’s predetermined fortune which all support the theme that one’s past is part of his/her present.

<table>
<tr><th>Question No.</th><th>Answer</th><th>Detailed Explanations</th></tr>
<tr><td>1</td><td>C</td><td>The degree of a term is the sum of the exponents of the variables in the term. The degrees of the terms are in the table below.
<table><tr><th>Term</th><th>Degree</th></tr><tr><td>$7x^9y$</td><td>10</td></tr><tr><td>$-9xy^8$</td><td>9</td></tr><tr><td>$2x^5y^6$</td><td>11</td></tr><tr><td>$-5xy$</td><td>2</td></tr></table>
So, $2x^5y^6$ has the highest degree.</td></tr>
<tr><td>2</td><td>B</td><td>A factor is a number that when multiplied gives another number or an expression. The given expression $36x^2+12x+24$ can be factored as $12(3x^2+x+2)$ which shows that 12 is a factor of the given expression.</td></tr>
<tr><td>3</td><td>B</td><td>The question asks us what the interest rate is in the expression $500(1.025)^7$. This expression uses the formula $P(1+r)^t$ where r is the annual interest rate as a decimal. Therefore, we will write an equation and solve for r:1.025=1+r;r=0.025. Since r is the interest rate as a decimal, we will multiply our answer by 100 and add the percent symbol. The interest rate is 0.025*100=2.5%.</td></tr>
<tr><td>4</td><td>C</td><td>The question asks us how many years money was accumulating interest the expression $1500(1.0355)^9$. This expression uses the formula $P(1+r)^t$ where P is the principal in the account, r is the annual interest rate as a decimal, and t is the time in years. Therefore, since the given expression contains an exponent of 9, the money was in the account for 9 years.</td></tr>
<tr><td>5</td><td>C</td><td>Student must set up an expression that subtracts the cost of the gloves from the total cost in order to represent the cost of the baseball bat; student must use the coefficient of 2 because there are two gloves being purchased.</td></tr>
<tr><td>6</td><td>C</td><td>The answer is C. Rainsford is objecting to General Zaroff's claim that hunting humans is a justifiable replacement for hunting the animals that no longer stand a chance against him</td></tr>
<tr><td>7</td><td>D</td><td>The answer is D. Unlimited success or perfection, as General Zaroff puts it, means that there is nothing to challenge the general anymore. Without a challenge, he is bored by the activity he loved most.</td></tr>
</table>

Question No.	Answer	Detailed Explanations
8	B	The answer is B. Having read Rainsford's book on hunting and declaring that it is his "a very great pleasure and honor" to meet him, Zaroff seems to admire Rainsford as an expert hunter.
9	B	The answer is B. Though the line contains the word "as," it is not used to indicate a comparison. The other lines each use "like" to indicate a comparison.
10	B	The answer is B. General Zaroff is being compared to a cat, and Rainsford is being compared to a mouse.

Question No.	Answer	Detailed Explanations
1	A	The rational exponent expression $x^{\frac{a}{b}}$ can be written as $x^{a.\frac{1}{b}}$ by separating the numerator and denominator in the exponent. Next, the expression $x^{a.\frac{1}{b}}$ can be written as $(x^a)^{\frac{1}{b}}$ using the power of a power exponent rule. Recall that $x^{\frac{1}{b}}$ is the equivalent of $\sqrt[b]{x}$, using the rational exponent rule. Now, we can put these rules together and get $y^{\frac{4}{5}} = y^{4.\frac{1}{5}} = (y^4)^{\frac{1}{5}} = \sqrt[5]{y^4}$.
2	A	The rational exponent expression $x^{\frac{a}{b}}$ can be written as $x^{a.\frac{1}{b}}$ by separating the numerator and denominator in the exponent. Next, the expression $x^{a.\frac{1}{b}}$ can be written as $(x^a)^{\frac{1}{b}}$ using the power of a power exponent rule. Recall that $x^{\frac{1}{b}}$ is the equivalent of $\sqrt[b]{x}$, using the rational exponent rule. Now, we can put these rules together and get $(xyz)^{\frac{3}{8}} = (xyz)^{3.\frac{1}{8}} = [(xyz)^3]^{\frac{1}{8}} = \sqrt[8]{(xyz)^3}$.
3	D	The expression $\sqrt[a]{x^b}$ can be rewritten as $(x^b)^{\frac{1}{a}}$. This expression can be changed using the power of a power exponent rule, which states that when we raise an expression with a power to a power, we multiply the two exponents. Therefore, $(x^b)^{\frac{1}{a}} = x^{b.\frac{1}{a}} = x^{\frac{b}{a}}$. Thus, $\sqrt[7]{x^5y^6} = (x^5y^6)^{\frac{1}{7}} = x^{\frac{5}{7}}y^{\frac{6}{7}}$.
4	A	The expression $\sqrt[a]{xb}$ can be rewritten as $(x^b)^{\frac{1}{a}}$. This expression can be changed using the power of a power exponent rule, which states that when we raise an expression with a power to a power, we multiply the two exponents. Therefore, $(x^b)^{\frac{1}{a}} = x^{b.\frac{1}{a}} = x^{\frac{b}{a}}$ Thus, $\sqrt[3]{x^3y^6z^9} = (x^3y^6z^9)^{\frac{1}{3}}$. Next, we use the power of a product rule and power of a power rule to simplify the expression: $(x^3y^6z^9)^{\frac{1}{3}}$ $= x^{3.\frac{1}{3}}y^{6.\frac{1}{3}}z^{9.\frac{1}{3}} = xy^2z^3$
5	A	The expression $\sqrt[a]{x^b}$ can be rewritten as $(x^b)^{\frac{1}{a}}$. This expression can be changed using the power of a power exponent rule, which states that when we raise an expression with a power to a power, we multiply the two exponents. Therefore, $(x^b)^{\frac{1}{a}} = x^{b.\frac{1}{a}} = x^{\frac{b}{a}}$. Thus, $\sqrt[4]{(abc)^9} = [(abc)^9]^{\frac{1}{4}} = (abc)^{9.\frac{1}{4}} = (abc)^{\frac{9}{4}}$.

Question No.	Answer	Detailed Explanations
6	D	The correct answer is D because all of those parts are in a traditional paragraph
7	C	The correct answer is C because the three points addressed in the thesis correspond to the three supporting body paragraphs that follow the introduction.
8	C	The correct answer is C because the thesis should be the last line of your introduction.
9	D	The correct answer is D because the topic sentence should link back to the thesis and explain how that paragraph supports the thesis.
10	A	The answer is A because it avoids using "I" and it is a complete thought that explains what the essay will be about.

Question No.	Answer	Detailed Explanations
1	A	To convert from degrees to radians, we multiply by $\frac{\pi}{180^0}$ To convert from radians to degrees, we multiply by $\frac{180^0}{\pi}$ Be sure that the 180^0 is in the denominator of the fraction, so that the degrees will cancel out in the problem.
2	B	To convert from degrees to radians, we multiply by $\frac{\pi}{180^0}$ To convert from radians to degrees, we multiply by $\frac{180^0}{\pi}$ Be sure that the 180^0 is in the denominator of the fraction, so that the degrees will cancel out in the problem. Reduce the fraction completely, but do not multiply by the value of π because we are asked to leave π in the answer so that the answer is exact (not a decimal approximation).
3	D	To convert from degrees to radians, we multiply by $\frac{\pi}{180^0}$ To convert from radians to degrees, we multiply by $\frac{180^0}{\pi}$ Be sure that the π is in the denominator so that it cancels out to leave degrees in the numerator of the answer.
4	A	To convert from degrees to radians, we multiply by $\frac{\pi}{180^0}$ To convert from radians to degrees, we multiply by $\frac{180^0}{\pi}$ First we need to find the measure of the angle. Since a circle is 360^0, when we rotate the terminal side by $\frac{1}{3}$ of the circle, that would give us 120^0 since 120 is $\frac{1}{3}$ of 360. Be sure that the 180s is in the denominator of the fraction, so that the degrees will cancel out in the problem. Reduce the fraction completely, but do not multiply by the value of π because we are asked to leave π in the answer so that the answer is exact (not a decimal approximation).

Question No.	Answer	Detailed Explanations
5	D	The circumference of a circle is calculated by multiplying 2π by the radius r, where 2π radians represents one complete circle. Thus The length of an arc is calculated as a portion of the circumference, so the formula for the length of an arc is the portion of the circle (called angle θ), times the radius, as long as angle θ is measured in radians. The given angle is measured in degrees, so convert the measure to radians by $36^0 . \frac{\pi}{180^0} = \frac{\pi}{5}$ radians. Thus, the length of the arc subtended by an angle of π/5 radians on a circle with a radius of 15 is $(\frac{\pi}{5}).15 = 3\pi$ inches.
6	B	The answer is B- the author's argument is that knowing our past will help us in our future, contrary to the beliefs of those who tell him to distance himself from the past.
7	A	The answer is A- the main idea is simplicity and how it enhances life.
8	B	The answer is B- although all of these claims are made at one point or another in the speech, the key word is overall, and the overall purpose is to remember those who died in the explosion.
9	A	The correct answer is A because these are connecting words in writing. Words such as "in a similar way" or "likewise" can be used to emphasize comparisons.
10	A	The correct answer is A because one must be able to prove the opinion to support the argument

Question No.	Answer	Detailed Explanations
1	A	When triangles are congruent, corresponding parts are also congruent. Since ∠A corresponds to ∠E,A=30°. Then all angles in a triangle must add up to180°, ∠C=26°.
2	B	Corresponding parts of congruent triangles are congruent. Since CD and NO are corresponding, their lengths must be equal. To find the length of CD we use the distance formula to find the distance between its endpoints. d= $\sqrt{(4-4)^2+(6-2)^2}$.
3	C	When triangles are congruent, corresponding parts must be congruent. ZX and CA are corresponding parts, so they must be congruent.
4	C	The hash marks on the sides of the triangle show congruence with the corresponding side of the triangles. The two congruent triangles can be identified by the pattern of the hash marks and the right angle in the triangles. Notice that on one side of the right angle, sides EF and HI are marked as congruent, and on the other side of the right angle, sides FD and IG are marked congruent. Then, sides ED and HG are marked congruent. Therefore, the congruency statement is ΔDEF ≅ ΔHGI.
5	A	The hash marks on the sides of the triangle show congruence with the corresponding side of the triangles. The two congruent triangles can be identified by the pattern of the hash marks and the appearance of the angles in the triangles. In the figure, ΔMNO is clearly a different shape than either of the other triangles. Notice that in the other triangles, sides GH and KL are marked as congruent, sides HI and LJ are marked congruent. Then, sides GI and KJ are marked congruent. Therefore, the congruency statement is ΔGHI ≅ ΔKLJ.
6	A	The answer is A, because infer means to deduce or conclude.
7	A	The answer is A. The sentence begins with the subordinating conjunction when, which means the end of the dependent clause needs a comma.

Question No.	Answer	Detailed Explanations
8	B	The answer is B. A pun is an intentional joke an author makes by exploiting the different possible meanings of a word.
9	B	The answer is B. Based on the context of Cindy working at a place where more women are employed than at her previous job, it's clear that homogeneous means "of the same kind," in this case, mostly men.
10	B	The answer is B. Based on the context of expecting equality, disparity means inequality.

Question No.	Answer	Detailed Explanations
1	A	A globe is in the shape of a sphere. To find the volume of a sphere, we need the radius of the sphere. Since the radius is ½ of the diameter, the radius is 40 mm. Use the formula for the volume of a sphere, $V = 4/3 \cdot \pi \cdot r^3$. Plug in the radius: V ‰ˆ 268,082 mm^3.
2	B	Remember the formula for perimeter is the sum of the length of the sides. For a square where all sides are equal: P=4s Substituting the value for perimeter 4 ft=4s 1 ft=s If each side measures 1 ft. The area will be: $A=s^2$ Substituting the value of s: $A=(1ft)^2=1ft^2$.
3	D	Remember the formula for circumference of a circle (the outline of a cylinder will be circular) C=πd Substitute the value of the circumference and π 6.28 meters= 3.14*d (6.28 meters/ 3.14)= d; 2 meters= d The radius is ½ the diameter or r=(½) d=(½)(2)=1 This value along with the height of the cylinder can now be used to find the volume $V=\pi r^2h=3.14*(1meter)^2*5$ meters=15.7 $meters^3$.
4	B	The trunk will be in the shape of a cylinder. Therefore $V=\pi r^2h$ Since the height is known, use the formula for circumference to find the radius C=πd 5π feet=πd 5 feet=d and the radius will be ½ that or 2.5 feet. Use this value in the Volume formula $V=\pi(2.5 feet)^2(10 feet)=62.5\pi feet^3$.

Question No.	Answer	Detailed Explanations
5	C	In this case find the area of the fence not the yard. So the length will be 20 ft and the height of 3 ft will serve as the other dimension. There will be two sections that have an area of $A=20\text{ ft} \times 3\text{ ft} = 60ft^2$ Doubling this: $60ft^2 \times 2=120ft^2$ The other side of the fence will have an area of $A=30\text{ ft} \times 3\text{ ft} = 90ft^2$ Adding these areas results in: $A=120ft^2+90ft^2=210ft^2$ Now divide this by the amount 1 gallon will cover: $210\text{ ft}^2 \div 50\text{ ft}^2$ per gallon = 4.2 gallons. John will need at least 5 gallons.
6	A	The correct answer is A. The purpose of this speech is to inform the citizens of America about his intentions as a leader and how he intends to work for the progress of the country.
7	D	The answer is D. The speech was made at the height of the Cold War, and the tone of the speech depicts strength, compassion, hope, and the belief that Americans working together can achieve growth and progress the country.
8	D	All of these are integral when accessing online sources.
9	A	The correct answer is A because that is how the story is presented – exposition is first, followed by rising action which leads to the climax, and then the story starts to fall and there is finally a resolution.
10	D	The answer is D because those are all the reasons we summarize.

Notes

What if I buy more than one Lumos Study Program?

Step 1 → **Visit the URL given below and login in to your parent account**

www.lumoslearning.com

Step 2 → Click on the horizontal lines (☰) in the top right-hand corner of the parent account and select **"My tedBooks"**

Place the Book Access Code and submit (See the first page for access code).

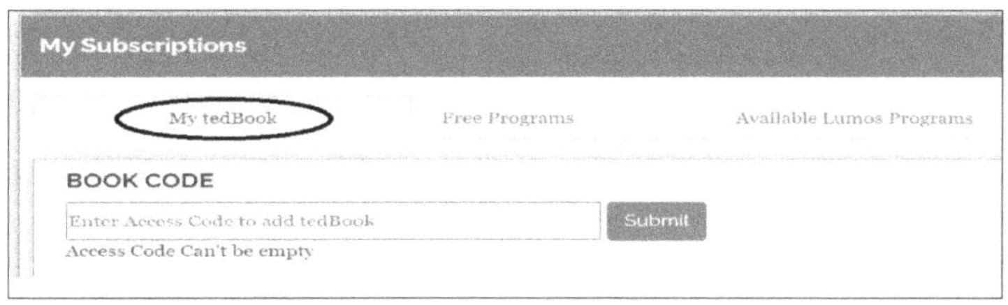

Step 3 → **Add the new book**

To add the new book for a registered student, choose the '**Student**' button and click on submit.

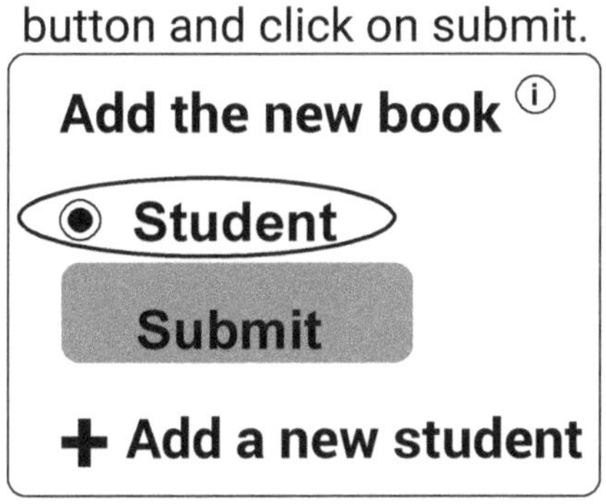

To add the new book for a new student, choose the '**Add New Student**' button and complete the student registration.

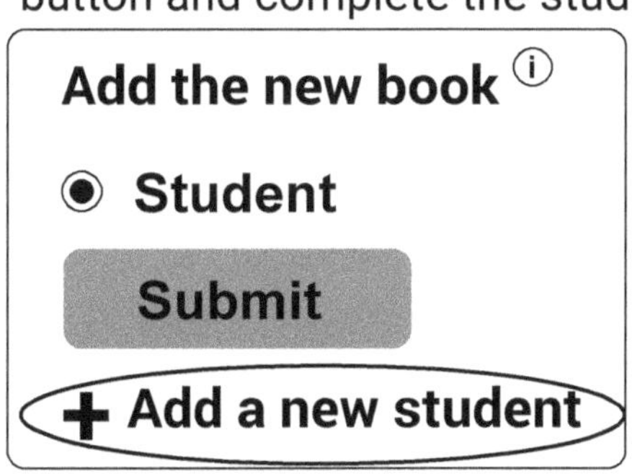

www.ingramcontent.com/pod-product-compliance
Lightning Source LLC
LaVergne TN
LVHW070851210926
848482LV00016B/912
* 9 7 9 8 8 9 7 1 3 1 0 8 2 *